SOTHIS PRESS

Sothis Press aims to completely reimagine the public domain book with an emphasis on universal access to captivatingly designed print editions. We believe that technology will enable every single book ever published to be perpetually accessible in beautiful print form. Sothis is beauty in ubiquity.

www.sothis.press
1 Waterhouse Square, London, ECIN 2ST

Secrets of the Stars

Inez N. McFee

SOTHIS PRESS

ISBN 978-1-0365-0811-1

This edition is published by
Sothis Press © 2023. The company
is registered in England and Wales.
The Company Number is 14900389.

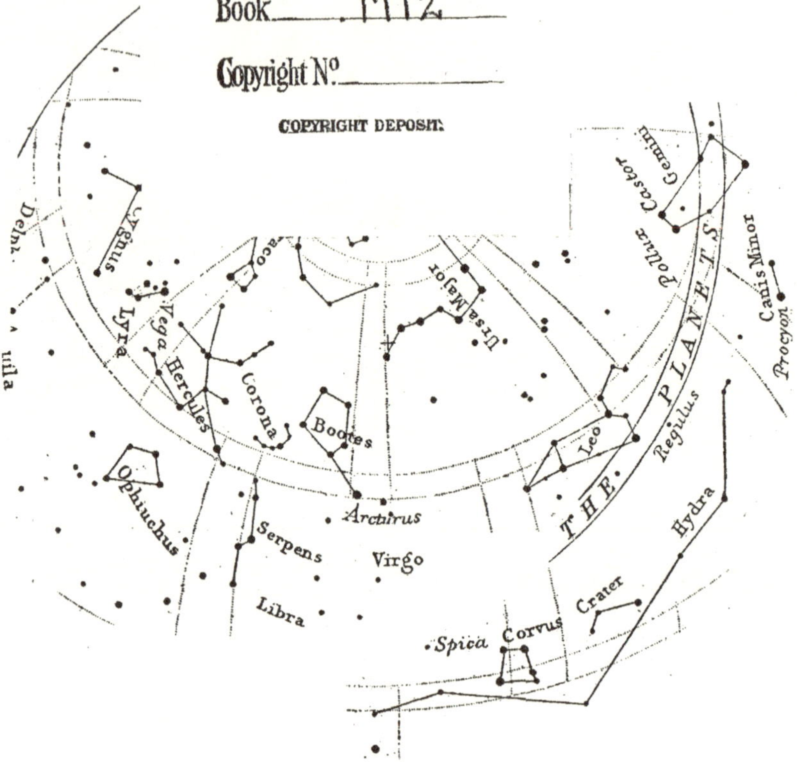

South

MAP OF THE STARS, 12 P.M. APRIL

SUMMER

North

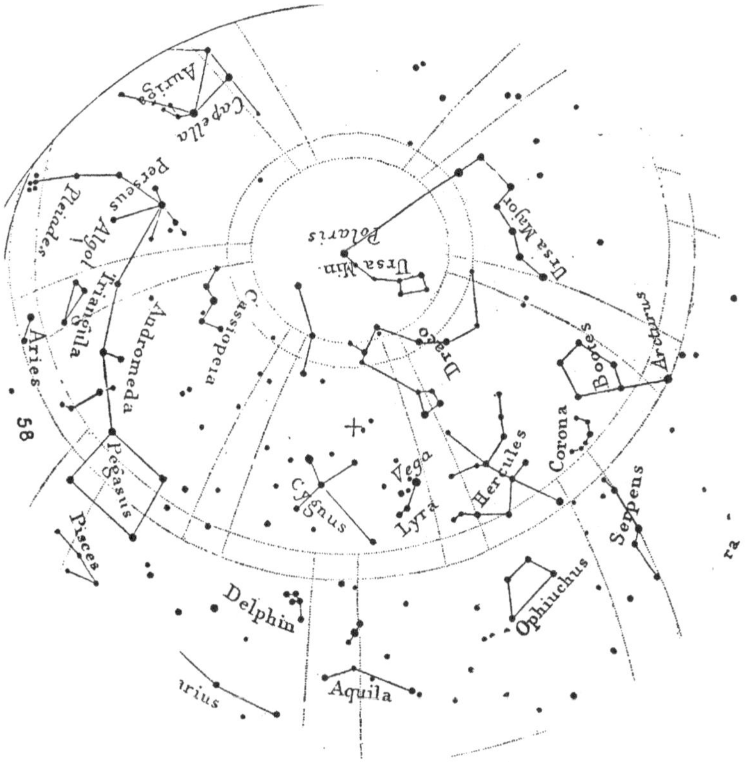

South

MAP OF THE STARS, 12 P.M., JULY

SECRETS OF THE STARS

BOOKS BY INEZ N. McFEE

SECRETS OF THE STARS
STORIES OF AMERICAN INVENTIONS
BOY HEROES IN FICTION
GIRL HEROINES IN FICTION
BOYS AND GIRLS OF MANY LANDS
LITTLE TALES OF COMMON THINGS
A PEEP AT THE FRONT
A TREASURY OF MYTHS
A TREASURY OF FLOWER STORIES

THOMAS Y. CROWELL COMPANY

SATURN, JULY 2, 1894

SECRETS OF THE STARS

BY

INEZ N. McFEE

Author of "Stories of American Inventions," "Little Tales of Common Things," etc.

ILLUSTRATED

NEW YORK

THOMAS Y. CROWELL COMPANY

PUBLISHERS

PREFACE

Emerson once said: "If the stars should appear one night in a thousand years, how would men believe and adore, and preserve for many generations the remembrance of the city of God which had been shown!"

But the stars are so common; we may look at them on every clear night, so the most of us pay them small heed. In all ages, however, there have been observing and thoughtful ones who have sought to read in the stars the answers to various questions about the Universe, until at the present time the world is dotted with observatories equipped with a great variety of instruments for studying the celestial bodies, and astronomy has come to be regarded as the golden chain between the Earth and the heavens. Indeed, so closely does it bind the two, that the one is mapped by means of the other. By the wise guidance of the stars, fleets and caravans are sent over wastes of sea and sand which would otherwise be trackless and impassable. By the stars are our clocks and watches timed, our lands surveyed, and the boundaries of our nations reckoned. In their twinkling faces are recorded not only the past and future of the comparatively small mud-ball which is our dwelling-place, but therein is also revealed the marvelous creation of the whole vast Universe. There is scarcely a science or an industrial activity of any kind that has not depended upon the revelations of starlight for its advancement. In truth, there is not a civilized person anywhere in the whole wide world whose life is not rendered more worth the living, or whose comfort is not affected, at least indirectly, by the work of astronomers and others who are applying the principles of this science to the practical affairs of daily life.

And yet so used are we to taking everything for granted, that, to most of us, the rising and setting of the Sun, the marvelous phases of the Moon, and the march of the glittering star hosts across the sky excites no wonder and inspires no awe in our breasts. We are dead to the wondrous truths, more glorious than any tale of mystery or romance, that are spread continually before our eyes in the great book of the sky—God's own clear page. Our painted American savages, the Arab in the desert, the simple, ignorant children of the race on the banks of the Nile, the wild men on the islands of the sea, have all been more observant than we as a general people are to-day! It is impossible to read the literature of modern times without stumbling upon references to star legends which have been handed down since time immemorial. They are the efforts of primitive man to understand and explain the wonders which they saw all about them. They constitute the poetry of the sky, which we shall certainly find delightful, even if our minds are so constituted that the drier, more scientific facts of astronomy do not appeal to us. Let us, then, rouse ourselves and at least make friends with the mysterious giant figures that people the blue dome of the sky! Once we learn to look upon these and to recognize them as the faces of our friends, we can but be interested in the countless things they have to tell us.

Nor is any apparatus really needed for our study of the stars, save that which Nature supplies—a good pair of sharp eyes. Remember, telescopes, while they may add greatly to the interest, were invented no longer than four hundred years ago, and all the planets and most of the stars that are now known had already been discovered and named. Likewise, too, countless of the most unfathomable mysteries of the skies have been discovered by no other aid than a clear eye and an inquiring mind. Have you these? Join hands, then, my readers, and let us go prospecting for the secrets of the vast suns of space —the boundless stars—which populate "the infinite meadows of Heaven."

The author gratefully acknowledges her indebtedness to the many published works on astronomy which are available, both for the beginner and advanced student in this fascinating subject. Authorities and works are cited in the text, but special thanks are due to Messrs. Cassell & Co. Ltd. for courteous permission to quote from Ball's "Starland."

I. N. McF.

Springdale, Ark.
July 1, 1922.

CONTENTS

"Ye stars! which are the poetry of heaven!
If in your bright leaves we would read the fate
Of men and empires—'tis to be forgiven
That in our aspirations to be great,
Our destinies o'erleap their mortal state
And claim a kindred with you; for ye are
A beauty and a mystery, and create
In us such love and reverence from afar,
That fortune, fame, power, life, have named themselves
a star."

—*Byron: Childe Harold's Pilgrimage.*

ILLUSTRATIONS AND DIAGRAMS

SECRETS OF THE STARS

I

WHAT THE STARS ARE

"Silently one by one in the infinite meadows of heaven
Blossomed the lovely stars, the forget-me-nots of the angels."

Perchance you may have stood out in the open only just last night and watched them! Perchance, too, you murmured softly to yourself, as you so often have:

"Twinkle twinkle, little star,
How I wonder what you are!"

But, do you not know? The stars are suns. Astronomers tell us that our Sun is a star, and that it is not nearly so bright, and by no means as large, as many of the stars which shine over our heads at night. But they are so far away that their splendid light seems to us but a mere twinkle.

The Sun is our nearest star—93,000,000 miles away. If we could fly thirty miles an hour, and should set out for the sun, not pausing for rest night or day, we should reach our destination in 345 years! Naturally the brightness of the Sun's glare, as it sweeps upon us by day, shuts out the glimmer of the more distant stars. But they are always in the sky, even though invisible to the naked eye. If we could stand upon one of the bright stars which we see in the heavens, and look toward our Sun, we would be surprised to find how dim and insignificant it looks. It would appear as a small star, if, indeed, we were able to see it at all.

"To the ancients," says Macpherson, in his *Romance of Modern Astronomy,* "the earth was the center of the Universe, fixed and immovable, the end and aim of the entire creation. Round the Earth revolved the Moon, the Sun, the planets, each in its own particular complicated pathway, and, farther away, the fixed stars, which they believed to be points of light fastened to the inside of a sphere. What lay beyond was outside the Universe. The whole Universe was supposed to be small in extent; its size was quite easily grasped by the mind of man. The Universe, too, in the opinion of the ancients, was created purely for the benefit of the Earth's inhabitants, the Sun to give light and heat, and the Moon to illuminate the nights, while the stars were regarded as convenient secondary light-givers in the absence of the Moon."

How much greater is our knowledge to-day! We know that the Earth is far from being the center of the Universe; it is not even the center of the planetary system to which it belongs. Indeed, instead of being the largest and most important body in the Universe, the earth is merely "the second-rate satellite of a second-rate star!" Nor are the dimensions of the Universe at all within the bounds of man's comprehension. The solar system alone is over 5,000 millions of miles in diameter.—What mind can conceive of this illimitable vastness?—And the solar system—that is our Sun with its planets, their satellites, and the comets—is a mere speck when compared with the greater system of the stars. Indeed, we are told that 9,250 solar systems of the size of ours could be contained in the space which isolates us from Alpha-Centauri, the nearest star. Truly the study of the stars—the science of astronomy—is the science of Infinity and Eternity. By means of the telescope and the spectroscope, we are led on an immeasurable journey away into boundless space where even Time itself is lost.

In the entire stellar system there are probably about 500,000,000 stars. Whether each of these stars is surrounded by attendants, like those which surround our sun, is impossible to determine; but it is altogether probable, and it aids us in realizing the extent and magnificence of the Universe in which we are situated. Between the earth and Alpha-Centauri (invisible only in southern latitudes) stretches an immense desert of space, twenty five billions of miles across. It is hard to grasp an idea of such a vast distance. Professor Ball says that it would take 300,000 years of ceaseless counting day and night even to count that number of miles. Again he says that all the cotton yarn ever spun in the world, joined in one long thread, would not reach to Centauri. Furthermore, he tells us that it takes light four years to come from this star to the earth.

But this is not all. When we look at the star now we see it as it was four years previously.

In fact, if the star were to go out altogether, we might still continue to see it twinkling for a period of four years longer, because a certain amount of light was on its way to us at the moment of extinction, and so long as that light keeps arriving here, so long shall we see the star showing as brightly as ever. When, therefore, you look at the thousands of stars in the sky to-night, there is not one that you see as it is now, but as it was years ago.

With the aid of our telescopes we can bring into view thousands of stars so far distant that their light must have been hundreds of years on its way to us. When we behold them, we do not see them as they are to-day, but as they were many, many years ago. Some, in fact, may be utterly extinct.

On the other hand, suppose there were astronomers living in these distant worlds. If they had telescopes powerful enough, they could witness events on our own planet—but they would see, not our busy life of to-day, but perchance some great scenes of the past—the Battle of Waterloo, Columbus discovering America, or the Crusades. Indeed, says one writer, "If we could view our own Earth from mirrors reflected in the stars, we might still see Moses crossing the Red Sea, or Adam and Eve being expelled from Eden!"

Thus it will be seen that if we had telescopes powerful enough to read and understand the stars, many of the secrets of the Universe would be such no longer. For example, if we could view the earth through the successive epochs of the past, all the geological problems that now puzzle us would be quickly solved. We should "be actually able to see those great animals whose fossil remains are treasured in our museums tramping about over the Earth's surface, splashing across its swamps, or swimming with broad flippers through its oceans." Indeed could we but interpret the revelations of starlight, we should have mirrored before us a story which for majesty, wonder, and sheer unexpectedness would stand unrivaled in the whole realm of literature.

At first thought it would seem wholly impossible to measure the distance of a star from the Earth. But astronomers tell us that the principle is no more difficult than those which frequently confront the land surveyor. The process merely involves taking certain angular measurements on the sky, and applying to them certain simple theorems well-known to the student of trigonometry. Right here is called into use a certain high-sounding astronomical term—*parallax*. In plain English the parallax of a distant point is the angle made by the crossing of two lines of vision directed one at each side of the object. Knowing this angle, together with the length of the base, the problem of distance is a simple one. It is not the rule, therefore, but its practical application which works the difficulty in obtaining star measurements. Using the diameter of the Earth's orbit, 186,000,000 miles, as a base line, the astronomer records his position and draws an imaginary line to the star he wishes to measure. Then he waits six months, while the obliging Earth changes his place in space, carrying him without any effort on his part to the opposite end of his base line, when he makes another observation. The shift or *displacement* of the star on the background of the heavens gives him the parallax angle which he has been working to secure.

The theory is good, but the trouble is that the displacement at best is exceedingly slight—"about equal to the apparent distance between the heads of two pins placed an inch apart and viewed from a distance of 180 miles!" Indeed, in the vast majority of instances, stars measured in this way show no apparent displacement, or at best give such a minute parallactic angle that it is impossible to secure trustworthy results. Fortunately, however, in this day and age we need no longer depend entirely on this method for star measurements. The perfection of an instrument called the heliometer for making angular measurements on the sky, together with the development of star-photography, and certain other more recent methods, have reduced star measurements to such a degree of accurate ease that we now have parallaxes based on the millionths of a second of arc. Indeed the farthermost object known in the heavens is a globular star-cluster at a distance of 1,260 *quadrillion* miles. Can one imagine anything like this enormous distance away into space? It is more than thirteen and one-half trillions times greater than the distance to the Sun. If one could embark on a cannon ball traveling half a mile per second, one would be eighty billion years reaching this goal. Again, a giant would need to take 13,300 billion strides as long as the distance from the Earth to the Sun to reach this far-off cluster.

From the very beginning of star measurement it was seen that the term *mile* conveyed little meaning when applied to star distance Therefore astronomers invented a new unit, the *light year,* for the distance traversed by light in one year. As light travels about 180,000 miles per second, it will be seen that the light year is well fitted by its stupendous magnitude for measuring the immense distances of stellar space. It also affords the best possible medium for our comprehension when used in comparison. For example, light crosses the diameter of the entire solar system in eight hours: yet it takes about four years to come from Alpha-Centauri. Light reaches us from 61 Cygni—our first star to be measured—at a distance of fifty-three billions of miles away, in about seven years; it comes from Sirius, the most brilliant star in the sky, fifty-eight billions of miles distant, in eight years. Light is a little over twenty-nine years in reaching us from Aldebaran, "the Bull's eye"; it takes forty-four years to come from the Pole star; and over two hundred years to travel from Arcturus, the brightest star in the northern constellation of Boötes, the Herdsman. How enormous must be this latter star which shines so brilliantly from such a vast distance! Compared with Arcturus, our Sun, great and splendid as he appears, is but a mere dot. Speaking of this enormous sun of space, Serviss says: "Imagine the earth and other planets constituting the solar system removed to Arcturus and set revolving round it in orbits of the same forms and sizes as those in which they circle about the Sun. Poor Mercury! For that little planet it would indeed be a jump from the frying pan into the fire, because as it rushed to perihelion, the point of its orbit nearest the Sun, Mercury would plunge more than 2,500,000 miles beneath the surface of the giant star. Venus and the Earth would melt like snowflakes at the mouth of a furnace. Even far-away Neptune would swelter in torrid heat."

All this but brings us to a fuller realization that indeed, "One star differeth from another star in glory." For not only are the stars to be found at all distances, but they vary in size, and frequently their marvelous immensity is even more overwhelming than their distance. We find it difficult to conceive of a body so vast as our Sun. A diameter a hundred times that of the Earth means little. Let us set up the familiar illustration of a foot-ball and a bird-shot side by side: now we have a good comparative idea of the size of the Sun and the Earth. If possible, try to imagine a gigantic orb equal to four thousand such suns as ours: this is Capella, a brilliant star midway between the Pole star and the constellation of Orion, the glory of our winter skies. Two other inconceivable orbs are Rigel and Betelgeuse, the two brilliant stars in Orion. The latter is one of the wonders of modern astronomy. It has been estimated as forty-three million times larger than our Sun! And Antares, a star in the constellation of Scorpion is said to be much larger than Betelguese.

More than this it has been averred that however startling the size of Antares may prove to be, there are other stars whose immensity may even surpass it!

"Infinite as the sands of the sea," thus the ancients styled the number of the stars in the heavens. To-day we know that this estimate is overdrawn, for were the stars in reality so numerous the whole heavens would be bright with a diffused light, and there would be no night. Indeed, you may be surprised to know that the stars that may be seen with the naked eye, counting over the whole range of the sky, are no more than 5,000 at best, while those which may be seen in any one locality will not number above 3,000. Thanks to the spectroscope we not only know of what the stars are made, but we are able to discern that they revolve in orbits and are moving through space at a high speed. Yet they are so far away from us that they seem to be fixed, and so we generally consider them, speaking of them as "fixed" stars to distinguish them from the planets or "wandering" stars. Thus, as one authority points out, "The constellation Orion preserves throughout the ages its well-known form. Similarly the Plow shines down on us to-day as it did on the kingdom of Israel and on the plains of Troy. So that for all practical purposes we are correct in speaking of the fixed stars. And yet, scientifically speaking, we are wrong. The stars are no more fixed than are the planets. Indeed, many of the stars are moving through space with a velocity far greater than the swiftest of the planets. . . . For example, 61 Cygni moves at the rate of thirty miles per second; while the bright star Arcturus has been calculated to have a velocity of no less than 376 miles per second. . . . But so distant are the stars, so deep are they sunk in the depths of space, that in the course of hundreds, even thousands of years, the casual star-gazer can detect no difference in their positions."[1]

Whither are all the heavenly bodies fleeing? This question troubled astronomers for ages, nor has it ever been entirely settled. We do know, however, that our own Sun, with his system of worlds, is moving along at the rate of eleven miles per second toward the constella-

[1] Macpherson.

tion of Lyra, aiming apparently for the star known as Delta-Lyræ. But, although this voyage has been in progress since the human race has been in existence, the eye of man has never yet been able to note any resulting displacement of the stars unaided. Moreover, it is certain that not less than 180,000 years must pass before our system, even moving at this impetuous speed, could possibly traverse the distance to the point now occupied by the nearest star. So that, while this mighty voyage through space is impressive, we have no cause whatever for alarm.

Another feat of the spectroscope is to show us that the stars are not all moving in the same direction. For example, the brilliant star Aldebaran is speeding away from our system at the astonishing rate of thirty miles per second. Some stars share "their proper motions with others." That is they move along at the same rate as some other neighboring star or stars. For instance, five of the seven stars which make up the familiar constellation of the Big Dipper are moving with the same velocity and in the same direction, and it is apparent that they form one star-system, even though they are separated from one another by billions of miles. Early astronomers concerned themselves a great deal over whether or not there was a central star round which all the other stars and systems centered, in the fashion displayed by our solar system. A certain German astronomer fixed upon the constellation of Perseus as the center of the stellar universe; another of his countrymen believed he had found conclusive evidence that the stars all centered around Alcyone, the chief star of the Pleiades. Neither of these theories ever found general acceptance, and today it is believed that there could be no one sun large and powerful enough to control the motions of all the other heavenly bodies.

Instead, astronomers generally have adopted the Milky Way as "the vast highroad to Jove's court." Here is assembled an infinite host of stars that, no doubt, have come in from the wide desert of illimitable space. Toward this milk-white road countless globular star-clusters and nebulæ seem to be rushing from either side. While other bodies—the mysterious spiral nebulæ—seem to be fleeing away! It is a baffling picture, and one concerning which all sorts of conjectures have been advanced. None seems clearer, however, than that voiced by Professor Garret P. Serviss, who sees the great Milky Way driving through space, "like a flat shining raft, built up of hundreds of millions of stars—our little Sun being lost among them—and drawing in from either side, and from distances of hundreds of quadrillions of miles, vast stellar organisms, of a globular shape, on which it feeds and grows, while from before it, like frightened flocks of strange winged creatures, hatched in the midst of the mysterious and boundless ether, flee the spiral nebulæ, speeding madly on—on—on."

Surely a strange, weird, and wondrously interesting romance is the story of the stars! Suppose that this very night, while we were attentively studying the heavens, there should suddenly spring into place, in a wide patch of blue, a bright and beautiful star! A star which no catalogues record, for the reason that it is a star which no one has ever seen before! And yet such an occurrence would be by no means unusual: any well-versed astronomer would be able to predict accurately the new arrival's course. For several nights it would grow in brilliancy, until perchance it ranked with the brightest stars in the heavens; then it would begin to wane and die down to the appearance of a faint star or perhaps disappear altogether. What does it mean? Whence came the brief visitor? At first it was supposed that these *temporary* stars, as they are called, rose from a violent collision between two heavenly bodies. The spectroscope, by analyzing the flame produced, not only showed this highly improbable, but suggested what is now universally accepted as the real solution, that two stars in their journey through space happen into a nebulous region, and thus flare up by reason of friction—a theory that is further supported by the fact that temporary stars seem nearly always to be involved with a nebulous haze. But the most amazing part of it is that this occurrence which we herald as the appearance of a new star is, in truth, only the message of a huge disaster, a mighty conflagration which took place hundreds, perhaps thousands of years ago, and has but now reached us on the wings of light.

Many instances of temporary stars are on

record, one of the earliest having been seen by Hipparchus, the famous Greek astronomer, in the year 134 B. C. One of the most notable heavenly conflagrations took place in May, 1866, when an ordinary telescopic star suddenly blazed forth with such intensity that in a few hours it increased its brilliance nearly a thousand fold. Careful spectroscopic study of its phenomena showed that the outburst was due to an explosion of hydrogen gas, and the incident caused no little uneasiness everywhere. "What would likely be the result, if a conflagration like that which took place on this remote sun were at any time to happen to our Sun?" queried an Edinburgh astronomer, thus voicing the fear of all. "Not only would all the various forms of life on Earth be utterly destroyed, but on all the members of our solar system there would be such a change effected, that if any life existed even on the remote Neptune it would at once be completely extinguished. Probably the life that existed on the whole system of worlds that circled round this distant star must have been annihilated, and as the heat and light of this star increased so very suddenly, there could have been given but short warning to the inhabitants of these worlds."

Temporary stars only blaze up once and then die down, but there is another class, called variable stars, which are quite as remarkable, and have given astronomers even more difficult hours to account for their peculiarity. These stars get brighter and brighter up to a certain point, then wane, only to brighten again, with a regularity that can be reckoned with the utmost accuracy. Some go through their course in three days, others take much longer, the maximum being 600 days. What is the explanation? A simple enough one, truly! Either these stars have a smaller dark companion which gets around into position every so often and shuts off the light, or they have a bright double, which in certain lines increases or decreases the brilliancy.

Still not half of the wonders of the stars have been told! We shall see as these pages progress how man has ever turned to the heavens to unravel the answers to much of Nature's puzzling phenomena. In the stars he has not only deciphered the past of our own world, but he has in a great measure been able to trace its future. For just as the buds, blossoms, and seeds of the plants in our gardens record the life history of their species, so do the different stages of star and planet life, from the various masses of luminous nebula to the dead worlds and dark stars, record the past and future of our own dwelling-place, the Earth. We are, as Kepler, one of the early astronomers, long ago pointed out, permitted in a measure to see the marvelously wonderful method by which the Creator called into being the magnificent Universe in which we live; to think as it were the thoughts of God after Him! We realize, too, as we ponder, a new depth and beauty in the words of the prophet Isaiah: "For My thoughts are not your thoughts, neither are your ways My ways, saith the Lord: for as the heavens are higher than the earth, so are My ways higher than your ways and My thoughts than your thoughts."

II

THE SUN, TO US THE GREATEST STAR OF ALL

THE early peoples of the world worshiped the Sun as the Lord of Day, the Fountain of Light, and the Bringer of all things good. But they were at a loss to account for him:

> Whence are thy beams, O Sun, thy everlasting light? Thou comest forth in thy awful beauty; the stars hide themselves in the sky; the moon, cold and pale, sinks in the western wave; but thou thyself movest alone.
>
> —*Ossian*

And whither? No one knew that the Earth was round, when Ossian wrote. They thought it was a great flat plain, extending in every direction. It seemed that the Sun daily traversed the sky only to plunge into the western sea! Indeed, there were those who fancied they heard the dreadful hissing noise when the glowing, red-hot orb dropped into the Atlantic! But here was the difficulty: how did it get around to the East, and rise fresh and hot as ever the next morning? Surely the plunge into the water should have quenched it! Some said that we had an entirely new sun each morning. The business of making the suns was supposed to be in the hands of the gods; all day and all night they worked with Titanic energy, and each succeeding morning saw a new orb launched forth on its journey.

This, however, seemed a waste of suns, and presently a new theory was evolved: Vulcan, the God of Labor, had the matter in charge. When the Sun dropped into the sea in the West, it was his business to rescue it and row it around the northern route to the East. Some of those who sat up all night to watch for indications of the progress of this journey, fancied they could trace the light of the glorious cargo along the northern horizon. Besides how else could one account for the long midsummer twilight? A tedious night's voyage poor old Vulcan had of it! Nor was there even a moment's respite. As soon as he reached the East, he must launch the Sun with terrific energy, so that it would not flag nor falter all the day, and then paddle back the way he had come, with unceasing industry, so as to be ready to catch the Sun in the evening, and be off again on his unending task!

Time has settled this problem, as well as that other great phenomenon, the change of seasons, which was even harder for the ancients to solve. Gradually people have come to understand these "miracles," and to take them as a matter of course. No longer do we feel awed when darkness settles down, and we have absolutely no fear concerning the return of day; neither are we troubled over the unending struggle between summer and winter. Likewise, too, we accept all the other beneficences of the Sun. But it is well for us to pause now and consider that without the Sun, life on the Earth would be impossible. To him we owe light, heat, power, food, water, in short, everything that we have, even the Earth itself. Perhaps we shall understand this latter statement better by illustration:

Draw down the shades, light the candles, and let us consider it while we have tea and muffins in the cozy radiance beside the fireplace. How cheerful the fire's ruddy glow!

Do you know that it is stored-up sunshine? "Why," you exclaim, "the fire is only coal." Yes, but what is coal? It is the fossil remains of giant trees, huge ferns, and luxuriant grasses and mosses which once grew by reason of the warm sunshine and soil-water. These muffins, too, were made from wheat grown by the Sun. Even the very power that ground the flour came from the Sun.

No matter whether ground in a mill turned by wind or by steam, the Sun did the work. For the Sun furnishes both wind and water. All the air currents on our earth arc duc to thc Sun in this wise: great tracts of land are warmed by the pleasant sunbeams. The air, in turn, is heated and rushes upward, while the cooler air slips in to fill its place. To do this, the air, of course, moves across the country. It is wind. And it is the Sun that has started it. All the water we use comes from the clouds, either as snow or rain. Even the water we have here in our teapot was floating far overhead, only a little while ago. Before that, perchance, it was just some tiny drops of water in the ocean. The Sun poured down his warm beams upon them, the drops turned to vapor, and climbed into the sky, there to unite in the cooler atmosphere with other rising drops and thus form a cloud, which floated our way. By and by the wind came rushing along, the vapor cooled still more and changed back to drops again, falling as rain, until presently the little cloud wept itself away. The drops seeped down deep into the soil, and at length found their way into the underground vein which empties into our well. Our tea is a product of sunshine. It came over sea in a ship built and run by sunshine. Look anywhere you may about the room: you can see nothing which does not owe its all to sunshine. Sunshine even grew and bleached this lovely white lunch cloth, and gave the pretty colors to your dresses.

Countless, indeed, are the marvels connected with the Sun, but of them all none is more astounding than its vast distance from us and its enormous size. Astronomers have determined that 93,000,000 miles of space stretches between us and the Lord of Day. But do you have any idea of this vast distance? Suppose you try to count 93,000,000. Professor Ball says the best way will be to let the clock do this: "How long will the clock have to tick before it has made as many ticks as there are miles between the Earth and the Sun? Every minute the clock, of course, makes sixty ticks, and in twenty-four hours the total number will reach 86,400. By dividing this into 93,000,000 you will find that more than 1,076 days, or nearly three years, will be required for the clock to perform the task."

Again, here is another illustration: It is nearly 25,000 miles round the world. The journey can be accomplished in sixty days. Traveling at this rate, one man might make 4,000 journeys around the world while another was covering the same distance between the Earth and the Sun. No man, of course, could accomplish this feat, even if such a journey were possible. He would be 600 years old when he reached his destination, even if he set out when a mere babe. As another instance, suppose that a train had started for the Sun, during the time of Cromwell, traveling at the rate of forty miles an hour, and stopping neither day nor night, it would not yet have reached its destination. It would require 265 years to complete the journey, and no man who started on the train could hope to be in at the finish. The journey must be ended by his great-great grandchildren.

No less amazing than the Sun's distance is his size. Approximately his diameter is 866,000 miles. This means that 109 globes the size of our Earth, set side by side across the face of the Sun, would not quite cover it. Again, comparing the volume of the Sun with that of the Earth, Professor Gregory gives the following illustration: "If we had a contract to build up this stupendous bulk, and were to deliver a load of the same size as the Earth every hour, the order could not be completed working night and day for 150 years."

Though the Sun is vastly greater than the Earth in volume, it weighs only 330,000 times as much. Thus the density of the Earth is about four times that of the Sun. The reason of this is that the Earth is a solid globe, while the Sun is a great glowing ball of incandescent gas. Nor is the force of gravity at the Sun's surface so great as his immense mass would seem to indicate: it is only twenty-seven and two-thirds times as great as gravity at the surface of the Earth. A body would fall vertically 444 feet in the first second. An athlete attempting stunts on the Sun would find his movements hampered by a bodily weight of 4,000 pounds. His running high jump, if possible at all, would not be over three inches. On the Sun, the pendulum of an ordinary clock would swing so rapidly that its movements could scarcely be counted.

The Sun is so constantly before us that we do not often stop to consider the immense amount of energy it is pouring forth every second. Astronomers have determined that the Sun is immensely hotter than a powerful electric arc-light which melts all known substances, and scientists have puzzled not a little to account for the work which this heat does in the vast regions of space; for only the two-billionth part of it reaches our Earth and is used to support the life of creatures here. On its surface the solar heat has been estimated at 18,000° Fahrenheit. How shall we realize this enormous heat? From even one square meter of this glowing surface enough heat is generated to supply 100,000 horse-power, continuously night and day. Further than this, Professor Young says: "If we could build up a solid column of ice from the Earth to the Sun two miles and a quarter in diameter, spanning the inconceivable abyss of 93,000,000 miles, and if the Sun should concentrate his power upon it, it would dissolve and melt, not in an hour, not in a minute, but in a single second; one swing of the pendulum and it would be water, seven more and it would be dissipated in vapor." Were the Sun as near to us as the Moon, his powerful rays would not only quickly turn all the ocean to vapor, but the solid Earth itself would be speedily melted.

How do astronomers know all this? By experiments with burning-glasses, mirrors, and arc-lights. The Sun's heat is determined by opposing it to an electric arc and measuring the heat which both produce at different distances. You may perhaps have started a camp-fire by means of the Sun's rays through a burning-glass; or perchance you may have seen a cannon covered by a burning glass in such a way that the Sun's rays touched it off at the noon hour. Sometimes one discovers black spots on leaves that have been caused by the drops of dew acting as little burning-glasses and concentrating the Sun's rays. It is an interesting experiment to make a lens out of ice, and burn a piece of paper by means of Sun rays which have passed through such a cold body. A mirror from an old automobile reflector is ideal for concentrating the Sun's rays, as its curved surface brings the beams to a direct focus. If the mirror is large enough an iron nail will burn like a match in the heat generated.

From experiments with large mirrors, we know that the heat of the Sun must be capable of melting all known metals, and glass and porcelain; for the Sun must be even hotter than the focus of the largest mirror we can construct, since a wall exposed to a blazing fire cannot be hotter than the fire itself. "With the Sun in the zenith," says Professor Todd, "his heat is powerful enough to melt annually a layer of ice on the earth nearly 200 feet in thickness."

The time is coming when the world's supply of bottled sunshine—wood and coal—will have disappeared; so that, within a century or two, new ways of furnishing heat and power must be devised, or the human race will perish of cold and hunger. Fortunately electricity and wind-power are at hand, but greater than either of these two sources is the unlimited might of the sea tides, which men have sought in vain to harness; and the direct heat of the Sun itself. Already solar motors have been invented—such as those by John Ericsson, the great Swedish engineer who built the *Monitor,* by Mouchot, Schuman, and others; the principal features to make them a success—cheapness and practicality—have not, however, been arrived at. Perhaps the key to these may be in the hands of some one who reads these words! Certain it is that the dependable solar motor is bound to come. Think what such a source of cheap power will do. Above all, it will make valuable countless acres of desert land, providing homes for thousands on profitable irrigated tracts. Any place in the tropics where the sun shines practically speaking throughout the year, and where ordinary fuel is very expensive, offers excellent opportunity for the solar motor. A small solar plant on the roof of the modern apartment house will not only light and heat it at the least possible cost, but it will cook the meals of the inmates, run the sewing-machine and the vacuum cleaner, and in short furnish power wherever power is needed.

One of the great mysteries in regard to the Sun's heat is its constancy. It shines to-day just as intensely as it did as far back as man has any record. How is this enormous heat kept up? Certainly it is not by the combustion of carbon. Had the Sun been composed of coal it must have burned out in less than five thousand years. And apparently the sun is in his prime! Scientists have decided that the only real plausible theory for the Sun's heat is that he is contracting upon himself. If this be true, he must have been vastly larger in the remote past, just as in the far distant future he will be very much reduced in size. It has been estimated that the burning of gases in a contraction of his diameter of six miles per century would fully account for his present heat. At this rate it has been calculated that the Sun will shrink to one-half of its present diameter in five million years. While certain types of life would disappear, enough heat and light would still be emitted for another five million years. But, after that period the glacial epoch would again dawn, and presently the Sun and his retinue would swing silently through space as cold, inert lumps of matter. That there are many types of such dead suns in the heavens is a fact well-known to astronomers.

No less remarkable than the heat of the Sun is his brightness. The intensity of sunlight, as it is called, has been calculated as 190,000 times that of a candle, 146 times that of a calcium light, and nearly four times brighter than the brightest electric arc-light; 600,000 full moons would be needed to equal the light of the Sun. Absorption by the Sun's own atmosphere not only reduces the amount of light which we receive, but changes its character. According to Langley, were it not for its atmosphere, the Sun would shine two or three times brighter than it now does, and with the bluish color of the arc-light.

Of course no one can look directly at the Sun, but by observing him with the aid of a telescope, or even through a darkened glass, we behold first a disc of yellow light—the *photosphere,* astronomers term it. Looking at this carefully, we may note that the brilliant surface is marked with dark spots. These are called sun-spots. But they are not permanent features of the Sun, like the mountains and craters are of the Moon. They are only temporary markings, here to-day perhaps and gone to-morrow. Sometimes the spots close up within a few minutes and fresh ones open elsewhere. Again the whole surface seemes mottled over in a curious way.

What does it all mean? Galileo and Scheiner, the astronomers who first discovered this phenomenon, could scarcely credit their senses. Surely the sun was too dazzlingly perfect to show defects, such as the sunspots seemed to be! At length, however, it was determined that the photosphere was merely the brilliant envelope or *light-sphere* of the sun, and that the dark spots were in truth great rents or holes in the glowing envelope, through which the core of the Sun was seen. Shortly, by means of the sunspots, astronomers were assured that the Sun, like the Earth, rotates on its axis. But "the day" of the Sun is twenty-five and one-

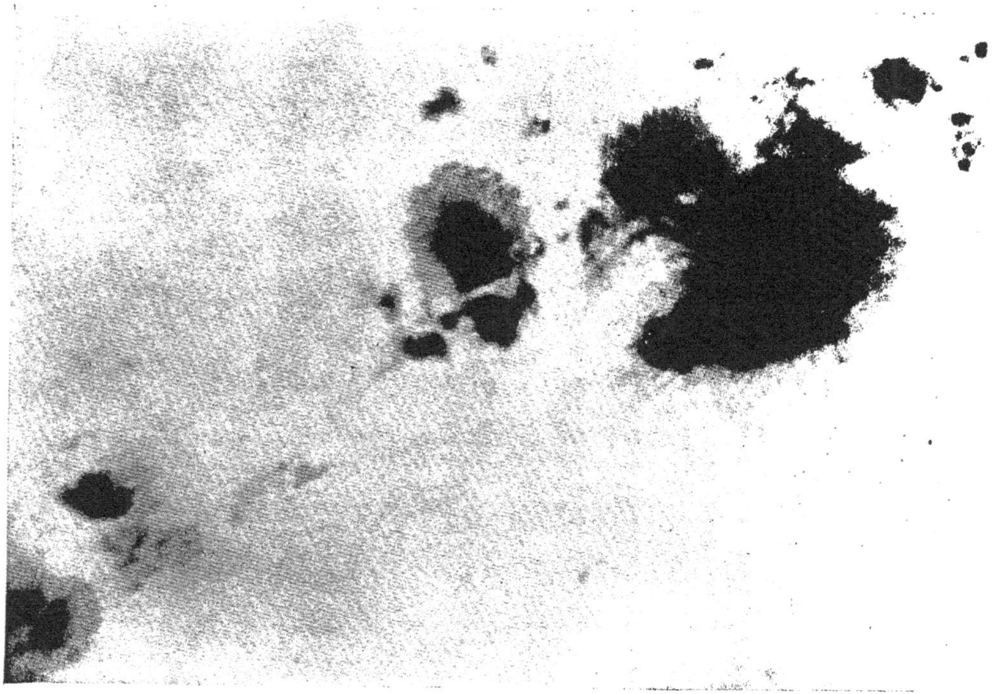

A GREAT SUN-SPOT, SEPTEMBER 11, 1898

quarter times longer than that of our planet. Another interesting fact is that the sunspots increase and decrease about every eleven years. Incredible as it may seem, the discovery of this *solar cycle,* as the increase and decrease of the sunspots is called, was made by a mere amateur in astronomy, whose only instrument was a hand telescope! In 1826, a German apothecary named Schwabe, who had been interesting himself in the heavens, commenced to count the number of spots on the sun as seen from day to day. After about twenty years had passed, he began to feel sure that there was a periodic increase and decrease of the sunspots, and by 1851 he had triumphantly proved the existence of the solar cycle—an important fact that had escaped the observation of all the noted astronomers. To-day it is known that when the sunspots are most numerous the aurora borealis is the most frequent and vivid; at this time, too, the presence of magnetic disturbances is noted in the fluctuations of the delicately mounted magnetic needles at Greenwich and in various other observatories. The largest sunspot on record covered about one thirty-fifth of the whole sun. This occurred in

1858. In February, 1892, another huge spot appeared, some 92,000 miles in length and 62, 000 miles wide. Closely connected with this huge gap in the photosphere were a number of smaller spots, in all making a rent that seventy bodies of the diameter of our Earth would have been required to patch.

Besides the sunspots, the telescope reveals another decidedly interesting feature on the bright surface of the Sun. This is the appearance of brighter streaks or ridges, called the *faculæ,* and wholly as irresponsible and notionate as the sunspots whose neighborhood they seem to frequent. Often it is impossible to sketch their form, so quickly do they appear and disappear. Some astronomers believe them to be elevated peaks of luminous matter extending thousands of miles above the gaseous plain. Some are evidently composed of clouds of incandescent calcium, an element which exists freely in the Sun. As a rule, sunspots are usually confined to the two zones above and below the solar equator, while the faculæ are often found well distributed over the Sun's surface, excepting in the "polar" regions.

For a large part of our knowledge of the Sun we have to thank the Moon. Occasionally this beneficent body gets between us and the Sun, producing an eclipse. Then, of course, we do not see the Sun, but we see something for the moment infinitely more interesting and beautiful. This is the wondrous spectacle of the *corona* and the *prominences.* The corona is the marvel of the astronomical world. Unfortunately it can be seen clearly only about seventy times in a century, though there is no doubt but that it is a regular part of the Sun—the third and final solar envelope, a halo glowing with a soft radiant silvery light. "Its shape," says Macpherson, "varies in sympathy with the eleven-year period, and it seems closely connected with electricity and magnetism. It streams out from the sun for millions of miles, becoming more and more rarified until it gradually fades into the ether of space."

In spite of its infrequency, various photographs of the corona have been taken. For it is possible for astronomers to figure when eclipses are due, the regions of their greatest visibility, and the length of their duration.

These dates are regularly published in the Nautical Almanacs, used by the English, French, German, and American governments, and naturally, at each occurrence, even though they must journey half-way around the world, representatives from every observatory are on hand, with a very definite idea of just what they want to accomplish in the few seconds, or perhaps moments, of the exhibition. Very carefully, you may be sure, are all the details rehearsed beforehand, time and again, that there may be the most accurate observation. In 1955 (Luzon) and 1973 (Sahara) will be two great total eclipses of seven and one-half minutes' duration, the longest for a thousand years.

The prominences, or red flames, which were first to be seen only at eclipse dates, are now, happily, to be observed almost any time. But they are not to be seen through even the most powerful of telescopes. You must call into service one of those ingenious oft-mentioned contrivances called a spectroscope—an instrument which is no more nor less than a special kind of a glass prism. By this means the prominences are shown to be tongues of glowing

SOLAR PROMINENCES, MAY 23, 1900

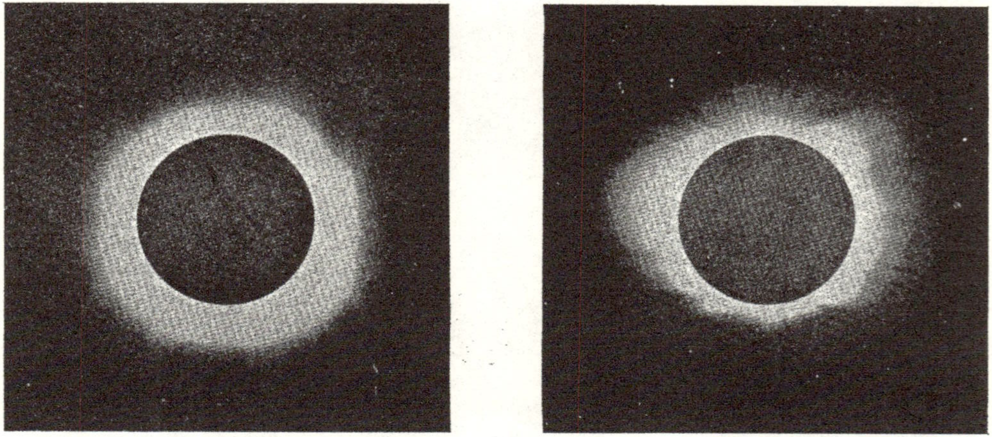

April 16, 1893　　　　May 28, 1900

THE SOLAR CORONA

fire shot forth with tremendous power from the *chromosphere*—an irregular rose-tinted fringe, varying from five to ten thousand miles in width, which completely encircles the sun, forming its second envelope. Many of these prominences are of enormous height, ranging from an average of 25,000 miles—about the circumference of the Earth—to 350,000 miles, nearly half of the Sun's diameter. Like the sunspots, the prominences increase and decrease every eleven years. Their variety of color shows them to be the gaseous vapor of various well-known metals, chiefly calcium (fiery-red) and iron (dark colored). Daily photographs of the Sun's prominences are taken in various observatories, but especially is this a feature at the Carnegie Solar Observatory in California. Two forms of prominences are known: brilliant eruptive jets and cloud-like wide-spreading shapes.

Few people who look up at the arc-lights which illuminate most of our city streets give any thought to how interesting and mysterious that light is. We know that it is the nearest thing possible to direct sunlight; in truth the arc-light is really the Sun's own imitation of himself. Thousands of years ago the Sun's energy was packed deep in the coal mines of the earth. To-day man has unearthed this treasure and turned it into carbon rods, which have again been led to produce the light and heat which the Sun stored up eons ago. By the aid of the Sun's own bottled energy, miraculous as it may seem, we have measured his heat. Through the medium of the multi-useful spectroscope we have watched various substances burn in the arc until their gaseous color is familiar on sight. By applying this same knowledge to the colors observed in the Sun's spectrum, we have been able to determine of what the Sun is made. So far about forty elements are known to exist in the Sun. Strangely enough, too, some elements were known in the day's bright orb long before they were found on our sphere: helium and coronium are two of these. Iron, hydrogen, calcium, nickel, and sodium are the most abundant of the Sun's elements.

Of what lies below the Sun's photosphere, in the center of his mighty orb, we can form no conception. According to one astronomer, "The pressure within the Sun is equally unconceivable. A cannon ball weighing 100 pounds on earth would weigh 2,700 pounds on the Sun. Thus a mighty conflict goes on unceasingly between imprisoned and expanding gases and vapors struggling to burst out, and massive pressures holding them down."

This, then, is our Sun, the center of the solar system, the great Fountain of Light, to which we on Earth owe all that we have. Besides our Earth there are seven planets of considerable size, and a whole host of insignificant little ones which depend upon the Sun. They all revolve about him, and derive their light and heat from his beams. In a good many ways these planets resemble the Earth. One of them, Venus, is about the same size. Mercury and Mars are much smaller, but Jupiter, Saturn, Uranus, and Neptune are a great deal larger. The nearest of these is millions of miles away, and naturally but little is known concerning them, though people have known of their existence for ages. The ancients named the days of the week from these seven celestial bodies.

You will perhaps like to make a drawing of our planetary system, in order to give yourself an idea of how it appears.

In order to draw the inner part of the solar system to scale, take a piece of blank paper at least ten inches square. In the center set up one leg of a drawing compass, open the other leg one inch, and describe a circle. The central dot, which can be enlarged is, of course, the Sun. The first orbit thus described is that of Mercury, and can be so marked, with "88 days" set down for the year, or time to complete the circle. The compass should next be opened one and three-quarters inches, from the center, and this circle will represent the orbit of Venus, 225 days. The third circle with a radius of two and one-half inches is that of our own Earth, 365 days. The fourth circle, with a radius of four inches, is Mars, 687 days. This completes the inner solar system. A very much larger piece of paper would be required to describe the orbits of the four outer planets.

Mercury is 37,000,000 miles from the Sun. The Earth is 93,000,000 miles. The great planets Jupiter, Saturn and Uranus are much farther. Jupiter takes twelve years to go around the Sun, Saturn twenty-nine and one-half years; Uranus eighty-four years; and Neptune, one hundred and sixty-five years. Jupiter, the largest of the planets, is twelve hundred times larger than the Earth. Of course these planets are so far away from the Sun that they get but little benefit from its light and heat. But, the gloom of their situation seems to matter but little, for it is highly improbable that any of these bodies can be inhabitated. We shall have more to say of these planets later.

"The most magnificent work of the Almighty," thus Schiaparelli defines our Sun, and so far as we on the Earth are concerned this is absolutely true. Indeed, as we view him in the light of all his marvelous greatness, we feel that the simple-minded ancients were not far wrong in bowing before him. "Surely," says Proctor, "if there is any object which men can properly take as an emblem of the power and goodness of Almighty God, it is the Sun."

III

MOTHER EARTH, A CHIP FROM THE SUN-STAR

OUR Earth, which the ancients thought held such an important position in the Universe, has now been proven only "a tiny grain of sand in the ocean of Infinity." But even so, it still remains the spot of the utmost interest to us, and no part of astronomy is more fascinating than that branch which answers the query, What is the Earth?

In its earliest stage, the Earth was probably whirled like a chip from the great mass of star stuff of which the Sun was being made. It was merely a chaotic nebula of gaseous formation, "without form and void," as the writer of the Book of Genesis sets forth. Then, "God said, 'Let there be light'; and there was light." But this light did not come from the Sun, says science; in its early days the Earth was self-luminous, light came from the slow contraction of the nebulous mass. By and by, as the heated mass went on whirling, keeping always in a certain path or orbit around the Sun, it slowed down a bit and began to cool off and a crust formed over its liquid surface, just as ice forms when water freezes. Then some of the gases formed clouds, and these were later cooled and condensed into water, which settled back in the hardened shallows and basins, made by the still cooling mass, to form seas and rivers. Others of the gases became air, and so, after some millions of years, the earth began to be a fit place for certain forms of life. But it was excessively moist yet, and the light was only the semi-twilight of deep shade, for the dense cloud masses shut off all the rays from the Sun, and the light and warmth were still supplied by the heated Earth. Ferns and plants of like nature sprang up and grew to the height of trees, covering the land with luxuriant vegetation. And still the contraction and vaporization went on over an Earth which was continuous and equal in climate everywhere. Not yet had the seasons begun; no outer Universe was visible.

And now God said: "Let there be lights in the firmament of the heaven to divide the day from the night; and let them be for signs, and for seasons, and for days, and years. And let them be for lights in the firmament of the heavens, to give light upon the earth; and it was so." For now the heavy clouds of vapor lifted, the skies cleared and the Sun shone in, bringing with him a rich heritage of beneficence. Now a new race of plants and animals came into being; the earth blossomed with loveliness, and the real history of the world began.

Almost in the beginning, too, men began to question one another. What is the Earth? And as we have seen, for long they fancied it one vast illimitable plain. Gradually, however, it was borne in upon them that the Earth was a great ball or sphere. Then came another baffling query; "If the Earth is a sphere, how is it supported in space?" The sky, they supposed, was a vast hollow sphere, or rim, above which dwelt the gods. Might there not be another rim below the Earth on which great pillars were placed to hold it in position? Some nations pictured a giant man who held the Earth on his shoulders; others fancied it upheld by various mythical animals of enormous stature.

"Some support, in the minds of the ancients, was absolutely necessary," says Macpherson. "The author of the Book of Job, however, had grasped the truth, for, writing of the power of the Creator, he says, 'He hangeth the Earth upon nothing.'" This we now know to be literally true.

Aristotle taught that the Earth was a globe suspended in space. It was, too, the center of the Universe, around which the Sun, Moon, and stars revolved. But presently a new problem arose to complicate matters. Attentive observers of the heavens now began to be aware of five bright stars which seemed to move in an irregular manner round the heavens, keeping close to the path which they fancied the Sun followed. However, there was good reason to suspect that these objects were not stars, and shortly the Greeks named them planets, this being their term for "wanderers." Each one of these planets seemed different from the others. One, the brightest of all, shone with such a soft, gentle light that they named it Venus after their goddess of love. Moreover they noticed that Venus never moved far from the Sun, and that they never saw it on a really dark sky. And presently they ascertained that Venus was sometimes visible as evening star, and again at another season of the year as morning star. Another planet, "the sparkling one," kept even closer to the Sun than Venus, and so they called it Mercury, the "messenger of the gods." Another of the planets, "the golden one," they called Jupiter, after their own marvelous deity, because it seemed mightier than the others, and instead of keeping close to the Sun, swept majestically round the entire heavens. One of a fiery color, which waxed exceedingly bright, then waned, they christened Mars, after their god of war. And last of all, and fainter than any, a dull yellowish, slow-moving planet, which crept around the heavens only once in thirty years, was called Saturn, after their god of time. But how were the paths of these planets to be explained?

Such a complicated theory as presently arose! to the effect that the Sun, Moon and planets moved in circles, and that the center of these circles revolved round the Earth in larger circles. It was fostered by Ptolemy, the

Egyptian astronomer, and for over 1,400 years it held sway. Nobody rightly understood it, but they had no better solution to offer. The theory of the spheres was an astronomical mystery too stupendous for any clear explanation. But at length rose one, Nicolaus Copernicus, who did not believe even what little he understood of the spherical theory. Nature always did things in the simplest, easiest manner possible, he argued, and surely there was a plain explanation for the system of the Universe.

Copernicus was only nineteen when he began to study the subject. He was brave enough to do his own thinking and to strike out in paths for himself, but he did not dare to publish the theory which he presently envolved—that the *Sun* was the center of the Universe, and that the Earth and the planets revolved around it, while the Moon revolved around the Earth. Surely no simpler accounting could be had, and the few great men of science who heard the Copernican theory longed to accept it. They felt perfectly assured of its truth. But those were days of bitter prejudice. The Roman

Catholic church had adopted Ptolemy's version; any other would be declared impious. Copernicus was an old man, seventy years of age, and on his death-bed, when his book came out proclaiming his solar theory. Bruno was burned alive for at once adopting it. Galileo and Kepler, the foremost astronomers, ran a gamut of persecutions, but the two bravely threw their whole souls into the effort to prove the questioned points which Copernicus had left. Shortly, then, the fact was established beyond a doubt that the Earth was merely a planet, a member of the Sun's great system of worlds.

But this theory only presented another problem. If the Earth was whirling around the Sun at the terrific speed of eighteen miles per second, as was claimed, and was also turning on its own axis once in twenty-four hours, how could objects stay on the Earth? How was it that the people did not fall off? Moreover, if we were rushing along at the great rate of over 1,000 miles per hour, why were we not conscious of motion?

It was Sir Isaac Newton who answered these

queries, in 1680. Most of our readers know the story of the falling apple which disclosed to his receptive mind the famous law of gravitation. By this law the Earth attracts everything toward its surface. We could not hurl ourselves off into space even if we ardently desired to do so. Mother Earth holds us by her power of attraction. Similarly this same law of gravitation keeps the Moon in its monthly orbit round the Earth, and prevents it from shooting off into space; also by this same principle the Earth and the other planets are kept in their path around the Sun. But how is it that the Moon is not drawn to the Earth, and the planets themselves pulled into the Sun and destroyed forever? What keeps them following accurate ellipses year after year?

Newton showed that this was due to certain other laws of gravitation and motion. For instance, the Earth attracts the Moon, and the Moon attracts the Earth; but the Earth is so much larger than the Moon that she overcomcs by mere size and forces the Moon into becoming a satellite. Again, when a body is once set in motion and is not acted upon by any force, it will move forward in a straight line, with unchanging speed, forever. Thus, the Moon in its journey around the Earth, and the planets in their course about the Sun, are each affected by the greater attraction of the larger body, but their own tendency to move in a straight line keeps them ever struggling outward. Hence, all the heavenly bodies in our solar system keep moving in an ellipse, or flattened circle, around and around their own great center of attraction. Astronomers call the motion of a planet on its axis the turning-motion, and its motion around the Sun the flying-motion. The reason we do not realize the enormous speed at which we are traveling, as the Earth rushes around its annual orbit of over 600,000,000 miles, is because the world is all traveling together, and with such ease and freedom from jolt that the motion is not noticed.

In swinging around its appointed orbit, the Earth does not stand up straight, but its axis points to the north star, causing what is known as the *inclination of the Earth's axis,* and this inclination and the motions of the Earth are the talismans which account for certain of the mysteries which puzzled the ancients—day and night, the seasons, twilight, the midnight sun, and the long polar night. The first of these phenomena, day and night, is caused by the Earth turning on its axis every twenty-four hours; thus the light of the sun strikes one-half of its surface at a time, making day on that side, while the other side, which is in shadow, has night. Owing to the fact that the Earth is surrounded by a great rim of atmosphere, daylight does not disappear the moment the Sun "sets," that is goes below the horizon; for his rays still strike the upper regions of the atmosphere, and thus we have a twilight gradually deepening into evening-tide and on to the total darkening of the sky. If the Earth stood straight, it is easy to see that the days and nights all over the world would be of the same length,—twelve hours each. Instead, the inclination of its axis is twenty-three and one-half degrees from the perpendicular. While at the poles we have one day and one night each six months in length.

For a few days at the summer solstice, the wondrous phenomenon of the Midnight Sun attracts many visitors to the northern parts of Norway, Russia, and Sweden.

It is hard to set down in cold print a picture of this glorious spectacle. The radiance of the Sun varies in intensity at different hours of the day, and on different days, depending upon the moisture or clarity of the atmosphere, presence of clouds, and other factors. One day it will be as red as orange, and so dull in radiance it can be beheld with the naked eye. At another time it burns with the vivid glow of live flame. Again there are days at a time when it has a bluish-white appearance, almost like the Moon.

But, the most curious effect of all is the motion itself. It does not "rise" or "set" as is apparent with us of southern climes. When one nears the North Pole there is no East or West or North. Every direction in which we look is due south. The Sun, therefore, apparently moves round and round us in a gigantic circle roughly parallel to the horizon. Gradually it descends in a slow spiral until it grazes the horizon, like a huge cartwheel turning constantly to hem us about. Next the cartwheel is cut in half—then only the upper segment shows. This disappears, and after a period of twilight, the long Arctic night sets in. The return of spring is just the reverse of this phenomenon—first the twilight, then a faint rim of light crawling round and round the edge of the landscape, making complete circles, until finally the orb of day ascends his throne again.

The rotation of the Earth on its axis furnishes a natural unit for the measurement of time. Two days are recognized: the *sidereal day* and the *mean solar day.* The first is the "day" used by astronomers, and is absolutely exact. It is the time required for the rotation of the Earth's axis as measured by the stars. It is calculated by choosing some particular star—Beta Cassiopeia, the right "pointer" at the top of the letter *W* formed by the constellation of Cassiopeia—is the "Greenwich of the sky"—and counting the exact time required for the star to move from the meridian round to the meridian again. But only an astronomer can manage the "juggling" necessary to keep "daylight" time by the sidereal clock. The reason is that, because of the motion of the Earth, the Sun appears to come to the meridian four minutes later each day by sidereal time. This means that the day measured by the Sun is four minutes longer than the day measured by the stars, and in a year the difference amounts to a whole day's time. Sidereal time would not do for the average unit of reckoning: "Sidereal noon," says one astronomer, "comes at all hours of the day and night during the progress of the year. Plainly, then, sidereal time is not a fit standard for regulating the affairs of ordinary life; for, while it would answer very well for a fortnight or so, the displacement of four minutes daily would in six months have all the world breakfasting after sunset, staying awake all through the night, and going to bed in the middle of the forenoon." Nor can the difficulty be altogether solved by taking any one solar day instead of the sidereal for measurement, for the Sun's apparent motion is beset with irregularities. Measured from meridian to meridian throughout the year, the solar days vary in lengths by many seconds. It is necessary, then, to find the average length of the total number of apparent solar days in the transit around the Sun, and this unit, called the *mean solar day,* is the common measure of time, by which all the clocks and watches in common use are regulated.

The rotation of the Earth, causing sunrise and sunset, also suggests a natural system of establishing directions. We have the cardinal points, east and west, north and south, and their divisions; we also have meridians and parallels, by which latitude and longitude are measured. Land surveys, by which boundary lines of real estate are marked, make constant reference to the cardinal points; boundary lines between states and nations are usually defined by meridians and parallels. Navigators have daily occasion to determine their position with respect to certain meridians and parallels to keep on in their desired route. In order that standard time may be kept throughout our country, the United States is divided into four time sections, each of which uses the mean solar time of its standard meridian. These meridians are exactly fifteen degrees, or one hour of time, apart. People traveling eastward across the United States advance their watches one hour on crossing each of these standard meridians; going westward the time must, of course, be turned backward one hour at each meridian. These facts give rise to a number of interesting problems, examples of which may be found in any text on arithmetic. Here is one selected at random: It is now 9 P. M. Tuesday, April 10, 1876, at Ann Arbor, Michigan; over what part of the Earth is it Tuesday, and what day of the week is it over the remainder of the globe? The answer is: It is Tuesday back to the east till we reach the point where it is *midnight* (i.e., as it is now 9 P. M., back 135°), and Tuesday west till we reach the 180th meridian. Between the 180th meridian and longitude 51° 9′ 12″ east it is Wednesday.

At midnight on Monday night at Greenwich, England, it is Monday all the way back to the east to the 180th meridian, and Tuesday all the way forward to the west to the same meridian. At the instant the midnight meridian coincides with the 180th, all the Earth has the same date day. In traveling eastward around the globe, we gain a whole day; and *vice-versa*, going westward, we lose a day. For convenience, there has been established in mid-ocean, closely following the 180th meridian from Greenwich, an imaginary line, called the *International Date Line*. Thus, were you traveling from California to China, and crossed the line on Monday, you would see posted about the ship notices that having crossed the International Date Line, the date is changed from Monday to Tuesday. On the return journey, should you cross this line on Monday, you would see a notice advising the change of the date to Sunday.

The chief effect of the Earth's revolution around the Sun, aided by the inclination of its axis, is the change of seasons, producing in turn spring, summer, autumn and winter, and the variations in the lengths of day and night. When the Earth is at *A*, March 21st, the spring or vernal equinox, it will be seen that the direct rays of the Sun fall upon the equator and that the great circle of illumination passes exactly through the poles. Day and night, therefore, are equal in all parts of the world. In the northern hemisphere are quickening Nature's coming signs of spring: "The melting of ice and snow, the gradual reviving of brown sods, the flowing of sap through branches apparently lifeless, the mist of foliage beginning to enshroud every twig until the whole country is enveloped in a soft haze to palest green and red, gray and yellow." In the southern hemisphere autumn has come, and unmistakable signs show that winter is on the wing. Continuing on in its orbit the Earth arrives at *B*, the summer solstice, June 21. Now, it will be noted, the entire region within the Arctic Circle is in daylight, while the region at the opposite end of the Earth is in darkness. Now, too, as the great circle of illumination passes beyond the North Pole, the days in the northern hemisphere constantly increase in length, and on June 22 is registered the longest day and the shortest night of the year. On the other hand, the opposite conditions prevail in the southern hemisphere. As the Earth traverses from *A* to *B* the days grow constantly shorter and the nights longer. The increased length of day in the northern hemisphere, of course, brings the conditions of summer; while the shortening day in the southern hemisphere brings winter's ice and chill. When the Sun reaches *C*, at the autumnal equinox, September 22, the Sun's rays again fall directly upon the equator and the great circle of illum-

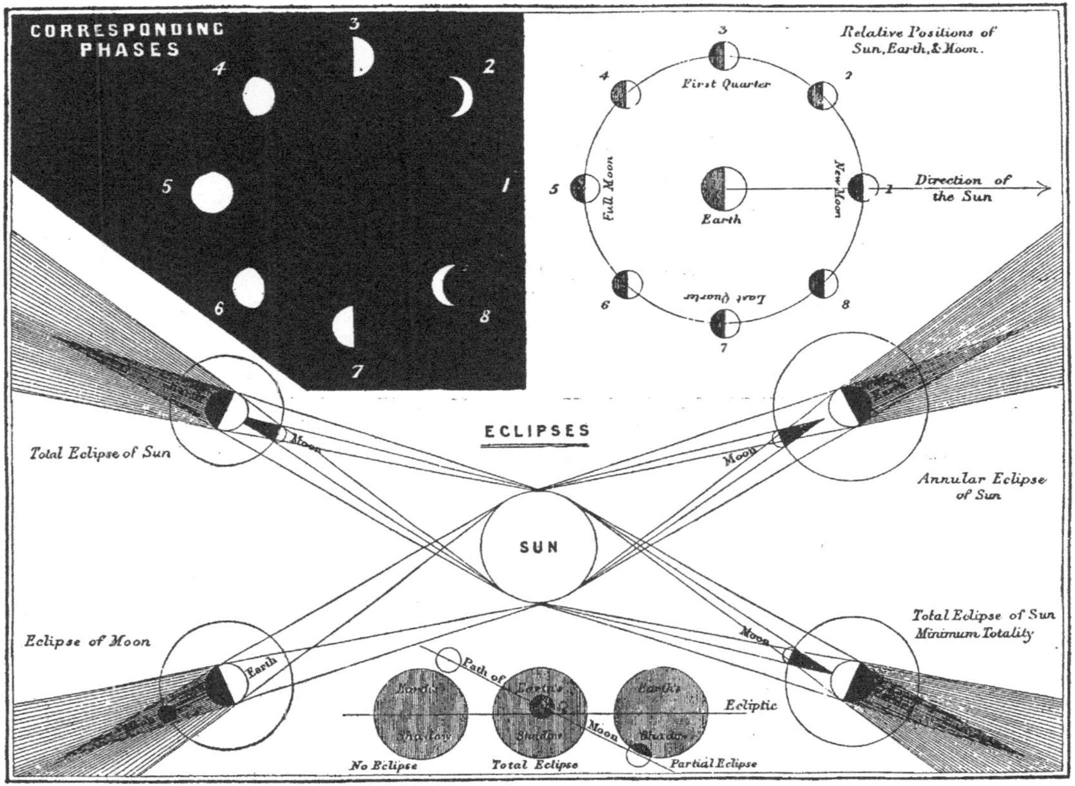
CORRESPONDING PHASES
Relative Positions of Sun, Earth, & Moon.
First Quarter
New Moon
Direction of the Sun
Earth
Full Moon
Last Quarter
ECLIPSES
SUN
Total Eclipse of Sun
Moon
Annular Eclipse of Sun
Eclipse of Moon
Earth
Total Eclipse of Sun Minimum Totality
Path of
Moon
Earth's Shadow
Ecliptic
No Eclipse
Total Eclipse
Partial Eclipse

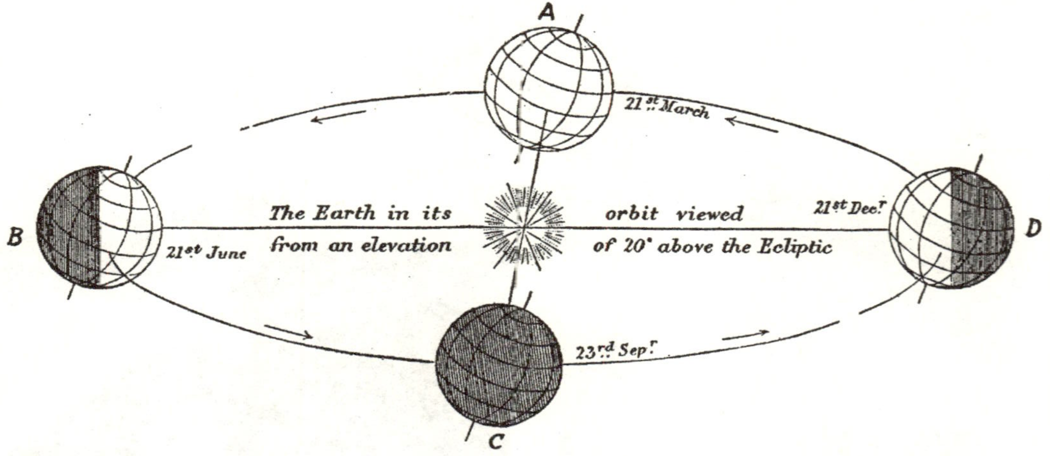

THE FOUR SEASONS

ination passes through the poles; thus once more the days and nights are equal. Now, in the northern hemisphere the summer season passes into autumn, and in the southlands spring succeeds winter. At *D*, the South Pole now enjoys the Midnight Sun, while the North Pole is enshrouded in the gloom of the Polar Night. Likewise, too, the southern hemisphere is having summer; while our part of the hemisphere has arrived at the winter solstice, or December 22, the date of the shortest day and the longest night. Now our part of the earth is tilted at its greatest angle away from the Sun. Now the trees stand dormant, their sap withdrawn far into their roots until the cold shall abate; the leaden skies swirl with snowflakes and all the Earth is hidden under a mantle of white. But now the world no longer trembles with fear lest these conditions prevail: we know that the Frost Giants can not long conquer, that presently the South Wind, "the summer-maker," will come, unlocking all fastnesses by his magic breath.

It is a well-known fact that cold is apt to actually increase for a month after the Sun has turned northward. Consider what it would mean if there were to be a permanent withdrawal of even a slight amount of the Sun's warmth: our Earth would freeze into perpetual winter. Again, a slight tilt of our axis might turn into Arctic regions those belts where the glory of summer now reigns in its turn. But nothing is more stable than the laws of the Universe; all changes of movement and direction are slow and gradual, and we need not apprehend any change from the familiar variation of seasons, at least not for any period of time within the appreciable grasp of man. In late January, when the Sun has long been climbing steadily on his northern route, we begin to take note of the lengthening days, but the Earth has arrived back to the vernal equinox ere the astronomers whys and wherefores pale before that mysterious thrill when—

> "Every clod feels a stir of might,
> An instinct within it which reaches and towers,
> And groping blindly above it for light,
> Climbs to a soul in the grass and flowers."

The Earth is made up of a vast number of elements, and its whole mass is estimated as six thousand millions of millions of millions of tons. This is an altogether incomprehensible term. Perhaps we shall convey a better idea if we state that the average weight of the earth is about 350 pounds per cubic foot. A cubic mile will, therefore, weigh in the neighborhood of 22,999,680,000 gross tons. The earth is not quite a perfect sphere, due no doubt to its centrifugal force when in a molten state. Its longest diameter is nearly 7,927 miles, and its shortest diameter nearly 7,900 miles; its average diameter then is 7,913 miles. Its average circumference is calculated as 24,880 miles, and the area of its surface is 197,000,000 of square miles. It has a solid crust, but there has been much difference of opinion as to the composition of its center. All early astronomers considered this to be a molten mass; later a theory was put forward that the interior of the earth was in all probability a gaseous mass, the existence of volcanoes and hot springs being cited as proof of this supposition.

One writer says: "In recent years the astronomer and physicist have collected evidence, which is as conclusive as such evidence can be, that the Earth is solid from center to surface, and even more rigid than a similar mass of steel. Lord Kelvin shows that if the Earth were a fluid surrounded by a crust, the action of the Moon would not cause tides in the ocean, but would merely tend to stretch out the entire Earth in the direction of the Moon, leaving the relative positions of the crust and the water unchanged."[1]

Measurement of the temperatures in wells and deep mines shows that the average increase of temperature downwards is about one degree for every sixty feet. Down twenty-five or one-hundred miles we should naturally expect to find a very high temperature; indeed we know that this is true from the melted lava that rises and escapes from volcanoes. Newcomb, however, says: "The matter of the interior of the Earth is kept solid by the enormous pressure. Thus, as we increase the temperature we have only to increase the pressure also to keep the material of the Earth solid. And thus it is, as we descend into the Earth, the increase of pressure more than keeps pace with the rise of

[1] "Astronomy for Everybody," Newcomb.

temperature, and thus keeps the whole mass solid." This view is now generally accepted.

Perhaps nothing about the Earth is more wonderful than its great transparent rim of atmosphere, reaching out into space to distances estimated as varying from forty-five to two-hundred miles. This atmosphere, or air, as we commonly term it, is composed of a mixture of gases, of which nitrogen and oxygen form the chief elements. Carbon-dioxide, which constitutes less than a thousandth part of the atmosphere, is yet an element of untold importance, since it is necessary for the structural growth of plants. Winds—air in motion—supply the waves and currents of the ocean. Soil, which is formed largely from decaying rocks and vegetable matter, depends upon the action of moist air. Rainfall, too, comes from moisture supplied by evaporation in the form of clouds, which are carried about by air currents. Atmospheric pressure and the contraction and expansion of air solves the problems of weather. Fire and heat, so necessary in all operations of mankind, could not exist without oxygen. All forms of life in the animal and vegetable world die in a few hours, if oxygen is withheld. In short, without the atmosphere, the Earth would soon stand as dry and barren as the Moon, a dead and burned-out world, whirling aimlessly on in the ceaseless path from which it can never escape so long as the Sun holds sway over his realms.

IV

SOMETHING ABOUT THE OTHER PLANETS

"If the comparison were not offensive to the Sun-god," says one of the well-versed astronomers, "we might say that he is like a spider at the center of his web. In the net of his attraction worlds are sustained. Relatively to his magnitude and might, the planets are but toys spinning round him." Yet we cannot realize the total extent of our solar system without a study of the marvelous array of planets and satellites, asteroids, comets, and meteors which revolve around our orb of day.

The planets naturally divide into two well-defined groups of four bodies each: the *inner* and the *outer* planets, separated by the ring or circle of a third group called the asteroids or minor planets. In the first division, ranging from thirty-six to one-hundred-forty-one millions of miles from the Sun, are Mercury, Venus, the

Earth, and Mars. Next comes the 900 or more asteroids, the largest of which is not quite 500 miles in diameter—less than the mean length of the state of Texas. In the outer group are the planets Jupiter, Saturn, Uranus, and Neptune, at a range varying from 482 to 2,789 millions of miles. To remember the arrangement of the planets in relation to the Sun, here is a sentence, the first letters of each word of which furnish the key to the name of the planet:

MEN VERY EARLY MADE JARS SERVE USEFUL NEEDS

It means little to say that any one of the planets is a certain number of million miles from the Sun. We need a concrete example: Here is one, as set forth by Macpherson: "If we take a nine-foot globe to represent the Sun, we may represent Mercury by a large pea at a distance of 127 yards; Venus by a one-inch ball at 235 yards; the Earth by a one-inch ball at 325 yards; Mars by a half-inch marble at 495 yards; the asteroids by 900 small seeds at distances from 676 to 1,385 yards. Jupiter will be represented by an eleven-inch globe a mile away; Saturn by a nine-inch globe one and three-

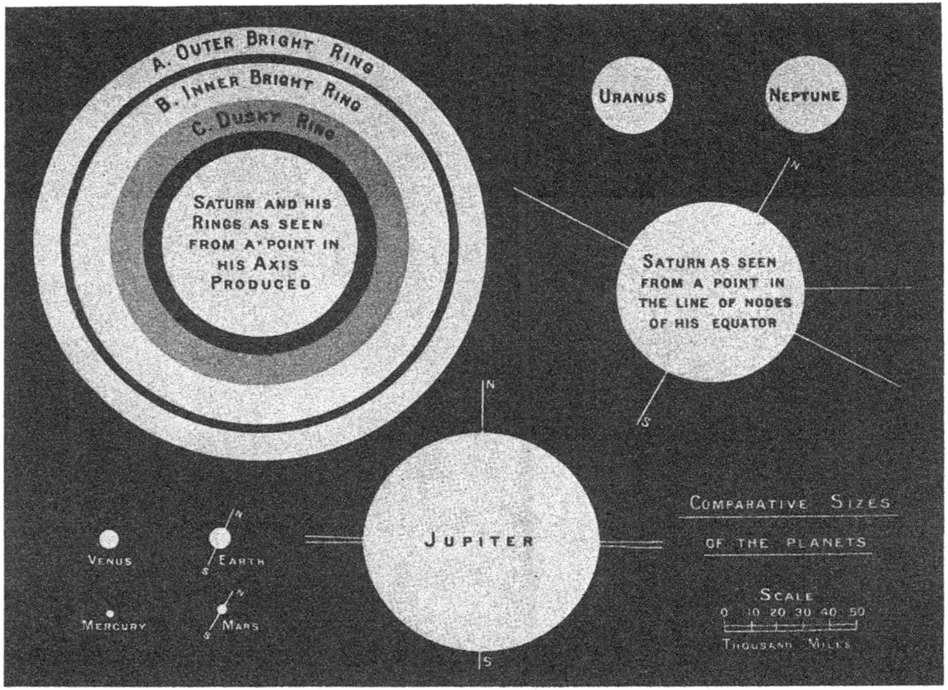
A. Outer Bright Ring
B. Inner Bright Ring
C. Dusky Ring
Saturn and his Rings as seen from a point in his Axis Produced
Uranus
Neptune
Saturn as seen from a point in the line of nodes of his equator
N
S
Jupiter
N
S
Venus
Earth
N
S
Mercury
Mars
N
S
Comparative Sizes of the Planets
Scale
0 10 20 30 40 50
Thousand Miles

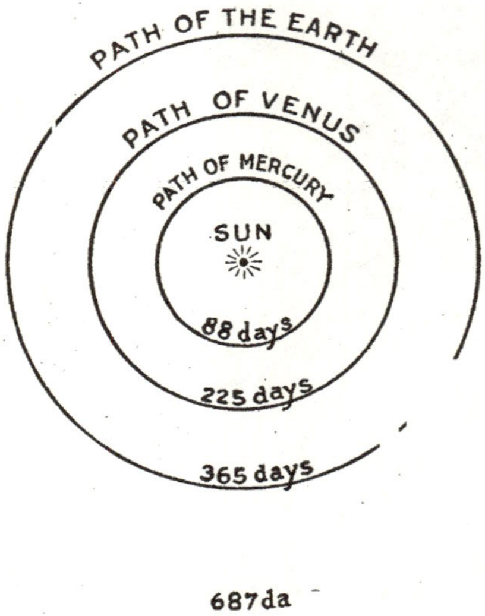

ORBITS OF THE INNER PLANETS

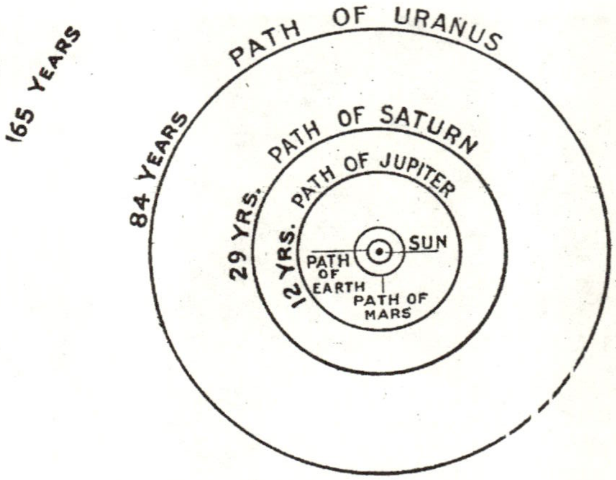

ORBITS OF THE OUTER PLANETS

Mean Distances from the Sun.

Mercury, 36 millions of miles
Venus, 67 millions of miles
Earth, 93 millions of miles
Mars, 141 millions of miles

Jupiter, 483 millions of miles
Saturn, 885 millions of miles
Uranus, 1,780 millions of miles
Neptune, 2,790 millions of miles

fourth miles away; Uranus by a fourteen-inch globe five and one-half miles away. On this scale we can represent the Moon as a pea moving in a circle at a distance of thirty inches from the ball, one inch in diameter, which represents the Earth."

Table Showing Actual Diameter of the Planets

Inner planets:	Mercury	3030	miles
	Venus	7799	"
	Earth	7913	"
	Mars	4230	"
The Asteroids, From 500 to 10 miles			
Outer planets:	Jupiter	92,164	miles
	Saturn	74,000	"
	Uranus	31,000	"
	Neptune	34,000	"

From the above it will be noted that the inner planets are very much less in size than those in the outer circles, and that the asteroids are often "chips" of such small magnitude as to be scarcely worth counting. Indeed, the three groups are completely different not only in size and distance but in physical condition. Moreover, certain of the planets are themselves the centers of particular little families of moons or satellites of their own. Mercury and Venus have none. Indeed only two of the inner planets have satellites. We all know the Earth's satellite, the Moon. Mars has two satellites, but both of these are smaller than our Moon. All the outer planets have an imposing circle of attendants. Jupiter has no less than nine moons, four large and five small. Saturn has ten attendants and a meteoric ring somewhat resembling small asteroids: Titan, the largest of these satellites, exceeds the planet Mercury in size, while the smallest are under thirty miles in diameter. Uranus has four satellites. Neptune, so far as wc know, has but one. All of the paths or orbits of the planets lie in the same plane as the Sun, but all do not journey about him at the same rate of speed, as their velocity depends on their nearness to power of attraction.

Table Showing Velocities of the Planets

Mercury	29	miles	per	second	2,505,000	miles	per	day
Venus	21	"	"	"	1,873,000	"	"	"
Earth	18	"	"	"	1,555,000	"	"	"
Jupiter	8	"	"	"	771,000	"	"	"
Saturn	6	"	"	"	536,000	"	"	"
Uranus	4	"	"	"	372,000	"	"	"
Neptune	3	"	"	"	268,000	"	"	"

Neptune, our farthest known world, travels, it will be noted, the slowest of all, three miles per second, 268,000 miles per day, and yet we see that this pace must indeed be a terrible velocity. No other known agent goes with anything like this swiftness. How inconceivable, then, must be the speed of our own world, which seems to us so still and immovable, but which really is carrying us onward through space at 1,555,000 miles per day; or, again, that of Mercury, which is rushing along 950,000 miles faster than we are! Flammarion gives this example: "A ball fired from a cannon leaves the mouth with a velocity of 1,312 feet per second, the terrestrial globe flies seventy-five times quicker, Mercury 117 times faster. This is a rapidity so stupendous that if two planets were to meet in their course the shock would be frightful; not only would they be shattered in pieces, both reduced to powder, but further, their motion being transformed into heat they would be suddenly raised to such a degree of temperature that they would disappear in vapor—everything, earth, stones, water, plants, inhabitants—and they would form an immense nebula." (The nucleus for a new world.) There is, however, absolutely no danger of a collision. The planets are all going the same way, and are separated by enormous distances. Mars and the Earth, at their nearest approach, are 40,000,000 miles distant, while Venus our nearest "neighbor" is 25,000,000 miles out in space.

Of course, as the planets travel at different speed, and are placed at different distances, their path or orbit around the Sun is traversed in different time. Thus, Mercury, being the nearest of all to the Sun, has a much smaller ellipse to travel; it therefore goes much more swiftly around its orbit, accomplishing its journey in comparatively short order.

Period of Planetary Revolution

Mercury	88 days		Jupiter	nearly 12	years
Venus	225	"	Saturn	" 30	"
Earth	365	"	Uranus	" 84	"
Mars	687	"	Neptune	" 165	"

It will be seen that our *year of time* as an element in the solar system is a very elastic quantity, depending entirely on the distance of a planet from the Sun. As one writer most cleverly points out: "A being who had lived only twenty-four terrestrial years would be a centenarian on Mercury, while the man of eighty-four on our planet would be an infant of one year according to the length of years on the planet Uranus."

Here is an experiment which will show the double motion of a planet revolving on its own axis and on its annual path around the sun. Take an ordinary dinner plate and half an egg shell. Moisten the rim of the plate and set the egg shell spinning on it. By tilting the plate a trifle, the egg shell will revolve in two direc-tions: first, on its own axis, and second, around the rim of the plate, which corresponds to the annual orbit of a planet on its course about the Sun.

No less remarkable than our unit of time in various parts of the solar system is that of our measure of weight. On the Earth the weight of an article is the force with which the Earth's mass attracts that body. Therefore, as the masses of the planets vary, a body must have different weights in accordance with the planet on which it is weighed. For example, a man who weighs 165 pounds on the Earth would weigh over two tons on the Sun. To express it differently, if you could take a one-pound weight and transfer it to the Sun, it would weigh twenty-seven pounds there. If you should try to raise your hat, you would find it weighed as much as a bucket of water on your head. In fact, you would hardly be able to lift your arms alone that far, as they would hang like lumps of lead against your body. And if you stooped to tie your shoe, you would be unable to lift your hundreds of pounds of weight again.

Macpherson, further explaining this question of weight, says that a man weighing twelve stone [1] on our world "would weigh twenty-eight stone on Jupiter, fourteen stone on Saturn, ten stone on Neptune, Uranus, and Venus. On Mars and Mercury the weight would be reduced to five stone, on the Moon to two, while on the asteroids it would come down to a few ounces. Let us suppose a man of twelve stone placed on the Moon. He would be amazed to find everything one-sixth as heavy as on the Earth. His own weight would be so diminished that he could jump over a house with as little

[1] England's legal measure of weight, about fourteen lbs.

effort as he could on Earth leap across a wayside ditch. Pulling out his watch he would feel practically no weight at all. A horseman who on earth would consider a five-barred gate a good jump, would on the Moon leap over a hayrick with the same amount of exertion. Suppose a man were playing cricket on the Moon. On Earth 100 yards is a very good throw; on the Moon one of 600 yards would be accomplished with the same amount of exertion. One able astronomer puts this lessened gravity very clearly: 'Football would show a striking development in lunar play; a good kick would not only send the ball over the cross-bar, but it would go soaring over the houses and perhaps drop in the next parish.'

"Next let us suppose our man of twelve stone weight transferred to one of the outer asteroids. Here his weight would be only a few ounces. He would feel as light as a feather. When he jumped into the air, he could easily clear a house or a tree. Football would be an impossibility. A good-sized kick would send the ball up so high, it would leave the asteroid forever—and become a little asteroid on its own account."

One interesting writer speaks of the planets as "The Sun's Kiddies." Sir Robert Ball, in considering the planets as the Sun's family, says: "Venus and the Earth may be considered the pair of twins, alike in size and weight. Mercury and Mars are the babies of the system. The big brothers are Jupiter and Saturn." According to one theory, the planets were originally parts of the Sun, and when they were thrown off by him, they continued to spin around on their own axis just as a ball continues its whirling motion after it has left the pitcher's hand. Reason has already been given why the planets go on spinning and why they are held in their regular path or orbit around the Sun. Another theory, known as the *Planetesimal Hypothesis,* gives still a different origin of the planets, as we shall see later on when we come to study the nebulae, and it is no less interesting.

Now let us find a way of telling the planets from the fixed stars. There are several ways of doing this:

(1) Look for the planets along the same general direction or path in which the Moon seems to be traveling.

(2) If a planet is above the horizon, it is the first object to be seen in the sky in the gathering darkness just after sundown, being styled as an "Evening Star." (The calendar will tell you what planets are evening stars on any given date.) Following this same rule of brightness, the planets are the last to be seen in the sky just before sun-up. They are the "Morning Stars." Rising early one morning, the writer had a splendid view of Venus, Mars, Jupiter, and Saturn—almost the sole remaining bodies of an especially bright and interesting night-time sky.

(3) None of the planets, except Mercury, twinkles, unless it happens to be near the horizon. They shine with a steady light, and their color names them: Mercury, pale ash; Venus, brilliant straw; Mars, reddish ochre; Jupiter, bright silver; Saturn, dull yellow; Uranus and Neptune, pale green. Five of the planets shine at about the brightness of a first magnitude star. Venus and Jupiter are a trifle the brightest; next Mars, Mercury, and Saturn in order. Uranus is about equal to a sixth magnitude star, and a few of the asteroids approach this. Neptune is

so far away that we glimpse him only as a star of the eighth magnitude, and then only through the telescope.

(4) When you locate a "suspect" look at it several times in the course of a few hours. Compare its position with reference to some fixed stars or group, and see if you can determine whether it is in motion. If so, you may feel sure that you have found a planet.

(5) Astronomers find that the very best way to locate the planets is by photography. This is especially true if the work be among the asteroids. Frequently time exposures are made of certain fields of the heavens. These exposures may last an hour or more. If in the field of vision there are stars only, they will be photographed as points of light. But if a small planet chances to be wandering along that way he will inevitably "make tracks" across the plate, and betray his presence by a continuous streak of light.

(6) Few people, however, are in position to photograph the sky. The next surest and best way is to use an almanac or calendar which tells you which planets can be seen at certain

times of the year, and in what part of the sky they may be found. Then with a "sky map," or a good star atlas, one can first locate the planet on paper, then verify its position in the sky.

MERCURY, THE SPARKLING ONE

Mercury is the nearest known planet to the Sun. Indeed, so close does this "messenger of the gods" follow the Orb of Day that he is never to be seen above the horizon more than about two hours after sunset or the same time before sunrise, and during this period he always seems to twinkle violently because his rays must break through the thickest part of the atmospheric rim. He is as hard to follow in his course as it is to trace his nimble namesake in the metallic world. But though Mercury is an elusive planet, he has been known to mankind for so many centuries, that no one now knows who his discoverer was. The ancient Greeks knew him well, and always referred to him as "the sparkling one." Three times during the year Mercury is morning star in the east, and three times he is evening star in the west. To determine the proper periods of the year to look for him, you must refer to the almanac.

Seldom can Mercury be found with the naked eye, nor does he yield any more readily to the telescope. He is usually so close to the Sun as to be overshadowed by the brilliance of that luminary. Like the other planets, Mercury is, of course, lighted by the Sun's rays, and he shows phases in the telescope just as the Moon does. One reason why Mercury is so hard to see is because we never see him against a dark background. When he does get on the opposite side of the heavens, the sun is between him and us. In other words, he is in *opposition,* as astronomers say when a planet lies in a straight line beyond the Sun. Now Mercury is the farthest from the Earth, and were it not for the Sun shutting out our view, we should see the planet with a fully illuminated disc, or "full Mercury." Coming out from his sojourn in the solar glare, Mercury does duty as *Evening Star;* then, as it comes nearer and nearer the Earth it begins to show crescents, until presently we have "new Mercury," and like our satellite at *new moon* it becomes invisible. Now the Earth,

Mercury and the Sun are again in a straight line. But this time they are at *conjunction,* because the planet is in the same part of the zodiac with the Earth. Coming on its journey Mercury soon appears as *Morning Star;* then again it shows crescents and finally disappears in the rays of the Sun to reappear again in its circuit as *Evening Star.*

In our almanacs *conjunction* is shown by the sign ☌ ; *opposition* by ☍. Suppose we turn to an almanac and note these signs ☌ ☿ ♂ indicated on a certain date: Can you interpret the meaning? It is the astronomer's way of saying briefly that on this date a conjunction of Mercury and Mars will take place; *i. e.* the two planets will be in a straight line and nearer to each other from our view than they will be again for a long time. Referring once more to the almanac we read ☍ ☿ ⊕: the interpretation is that on the date these signs are marked there will be an opposition between Mercury and the Earth; *i. e.* Mercury will be beyond the Sun and therefore hidden from view.

Mercury, you remember, is the swiftest of all the planets, rushing along at the stupendous speed of twenty-nine miles per second, but it only makes one turn on its axis in its journey round the Sun. Thus it turns the same side always to the Sun just as our Moon does to the Earth. One side of Mercury is always flooded with perpetual sunshine, while the other is black as night. One side is baking hot, while the other is locked in icy throes. Small wonder that its surface seems cracked in all directions: "a geography in black and white." Owing to the irregular motion of the planet, and its slightly varying velocity, due to an erratic deviation in its orbit, there is a small zone on Mercury where the sun rises and sets. "In fact," says Mr. Gore, "an inhabitant living on the planet's equator would have forty-four days of sunshine and forty-four days of night and twilight. A little farther in on the dark side there would be perpetual twilight; and farther in still, eternal night would reign. Owing to the low altitude attained by the Sun near the bounding line, its intense heat and light would of course be much mitigated, so that probably this region of the planet's surface may be comparable with the temperate zones of the Earth."

It is the general opinion of astronomers, however, that there are no dwellers on Mercury. At least, there can be no life there such as we know. Its surface is too rugged and mountainous, and it is extremely doubtful whether there is any atmospheric envelope. Of course there is no air, neither is there any water, but these points are debatable. "If there are inhabitants of Mercury," says one writer, "they must from the dark side of their world, obtain magnificent views of the outer Universe. Venus and the Earth will shine with a glorious radiance, fully illuminated. The Earth and Moon seen from Mercury form a fine double star."

Mercury is the smallest and lightest of the planets; in fact, it would take twenty-five planets of his size to weigh as much as the Earth. Perhaps you wonder how we know this. An astronomer must be a wizard at calculation, and also what Edison terms a "good guesser." Ordinarily the easiest way of determining the size of a planet is by the attraction it has for its satellites: the problem is one of falling bodies, intricate enough, but also very exact, as proven by other rules. But Mercury has no satellite.

Other methods, then, had to be resorted to, and the result of one of these, as told by Sir Robert Ball, in one of the series of lectures which he gave in London more than thirty years ago, is wonderfully interesting and amusing:

"There was once," said he, "and there is still, a little comet which flits about the sky; we call it after the name of its discoverer, Encke. There are sometimes splendid comets which everyone can see—we will talk about these afterwards—but Encke is not such a one. It is very faint and delicate, but astronomers are interested in it, and they always look out for it with their telescopes; indeed, they could not see the poor little thing without them. Encke goes for long journeys through space—so far that it becomes quite invisible, and remains out of sight for two or three years. All this time it is tearing along at a tremendous speed. If you were to take a ride on the comet, it would whirl you along far more swiftly than if you were sitting on a cannon-ball. When the comet has reached the end of its journey, then it turns round and returns by a different road, until at last it comes near enough to show itself.

Astronomers give it all the welcome they can, but it won't remain; sometimes it will hardly stay long enough for us to observe that it has come at all, and sometimes it is so thin and worn after all its wanderings that we are hardly able to see it. The comet never takes any rest; even during its brief visits to us it is scampering along all the time, and then again it darts off, gradually to sink into the depths of space, whither even our best telescope cannot follow it. No more is there to be seen of Encke for another three years, when again he will come back for a while. . . . And he is then so shy that usually very few catch a glimpse of him."

Now it seems, to tell the story briefly, that an astronomer and a mathematician were great friends, and by giving heed to what the astronomer told him about Encke, the mathematician, who was what boys would term "a whirlwind at math," shortly figured out a route and a time-table for Encke. Moreover, it was absolutely correct, as subsequent visits of Encke proved. Thus, all went happily for some years, and then lo! Encke failed to appear according to schedule.

What was the reason? The two friends came well-nigh quarreling. But the mathematician was certain his figures were correct. Something had evidently happened to Encke. And shortly the truth of this was apparent. Encke turned up all right, but not exactly in the place he should have appeared; moreover there were several indications that he was considerably jarred and upset. Very plainly he had been mixed up in a tussle somewhere!

The mathematician looked over his figures and drawings. In one place along the route Encke came rather close to Mercury's orbit. "See what Mercury has been doing lately," shouted the mathematician, exultantly.

The astronomer consulted his friends, and to his amazement found that erratic little Mercury had also been showing every indication of having been up to something, and the problem was solved. The mathematician, being also "a good guesser," had been able to lay his finger on the trouble. He knew that Encke had come near enough to Mercury to be drawn by his attraction. In short, the law of gravitation had very nearly worked chaos for Encke! The astronomer was for dismissing the matter with perfect satisfaction, but not so the mathematician.

"Wait a bit," he said. "It is the part of wisdom to benefit by such mishaps. Let us see what actually happened to Encke. Let us measure the distance between the place where Encke is, and the place where he ought to have been."

Subsequently the mathematician proved that Encke's delay was a measure of the mass of the planet, and triumphantly produced the figures which totaled the weight of Mercury's globe.

VENUS, THE EVENING STAR

Venus at her best is so much brighter than any of the stars or planets—twenty times brighter in fact—that one cannot help recognizing her on sight. But you must choose the right time to look for her: early in the morning in the east before the Sun is up, or in the west just after the Sun has gone down. Nor can Venus be seen every day. You must consult the almanac for the time to look for her, as, like Mercury, she is invisible the greater part of the year. "When first seen as an evening star," says Ball, "Venus will often be like the Moon at the quarter, and then it will pass to crescent shape. Then the crescent becomes gradually thinner, and next will follow a brief period of invisibility before the appearance of Venus as the morning star."

Like Mercury, Venus makes but one rotation on her axis during the 225 days it takes her to go around the Sun, and she also keeps the same face toward the orb of day. Hence there are weeks at a time when she cannot be seen because of the Sun outshining her when she is between us and the Sun, or by being invisible because she is at the other side of the Sun. Both Mercury and Venus attain their greatest brilliance when close to conjunction. For many days near this time Venus is visible in the clear blue even at mid-day, and she is often so dazzling in the early evening as to give credence to marvelous tales. In 1887-88, so wondrous was her spectacle that she was hailed as the "Star of Bethlehem." In 1897, she shone forth with such splendor that a facetious newspaper reporter set adrift a tale citing her as an immense electric light, attached to a balloon, which had been sent up from the Edison laboratory to apprise the world of the new light which had just been invented.

Unlike Mercury, Venus runs nearly true on her orbit; hence one-half of her globe is always in the bright sunshine, while the other half is in darkness. Moreover, so dazzling is her light, which is, of course, simply reflected from the Sun's own gorgeousness, that we can not study her, even when she faces us plainly. Little, therefore, is known about the real surface of Venus. She seems to be rugged and mountainous, and in size very nearly approaches that of the Earth. More than one astronomer has figured that Venus might be inhabited by mortals somewhat resembling ourselves. While she is nearer the Sun than we are, this does not necessarily mean a hotter climate. Other conditions of atmosphere might offset this.

Something over a year ago, Marconi startled the world by the announcement that he had received wireless signals which led him to believe that we were being signaled by another planet. "They steal in at our stations at all seasons," said he. "We do not get the signals unless we establish a minimum of sixty-five mile wave lengths. Sometimes we hear these planetary or interplanetary sounds twenty or thirty minutes after sending out a long wave. They do not interrupt traffic, but when they occur they are very persistent. The most familiar signal received is curiously musical. It comes in the form of three short raps, which may be interpreted as the Morse letter 'S,' but there are other sounds which may stand for other letters."

Australia immediately corroborated Marconi's statement. Highly skilled and experienced operators at Sidney reported frequent repetitions of two dashes, representing the letter M. They came in on wave lengths of 80,000 to 120,000 meters. Such wave lengths are not in common use at any wireless station of the earth.

Thomas A. Edison, commenting on the possibility of such communications, said: "If we are to accept the theory of Mr. Marconi that these signals are being sent out by inhabitants of other planets, we must at once accept with it the theory of their advanced development. Either they are our intellectual equals or our superiors. It would be stupid for us to assume that we have a corner on all the intelligence in the Universe."

If either Mercury or Venus at conjunction is near the node of the orbit, that is near one of the two points where the orbit intersects the ecliptic, or path of the Sun, the planet can be seen to pass across the Sun like a round black spot. This is called a transit. There are about thirteen transits of Mercury in a century, the shortest interval being three and one-half years, and the longest thirteen. As the Earth is near the nodes of Mercury's orbit in early May and again in November, it is in these months only that the transits can happen. Because Mercury's least distance from the Sun is at the November period, there are far more transits of Mercury in this month than in May. The Venus transits occur in pairs, eight years apart, during the months of June and December respectively, at intervals of 121½ years and 105½ years alternately, a June pair in one century being always followed by a December pair in the next. Both Mercury and Venus at transit are then nearest the earth, and their apparent motion is westerly or *retrograde*. (Tending backward.) For this reason a transit always begins on the east side of the Sun. The last transit of Venus was in 1882. Therefore no one of our readers will ever be able to see this wonderful phenomenon, which has always attracted the world of scientists, not so much for its beauty or its singularity, but because of the valuable information it affords. By taking certain measurements from both sides of the Earth of the displacement produced in the transit of Venus, mathematicians are able to calculate the distance of this planet, and knowing this they can also not only compute the distance of the Sun, but the distance and size of the other planets as well. This method of computing the Sun's distance is, however, only of historical interest. Other and more accurate methods have displaced it.

MARS, THE RED PLANET

Mars has probably received more attention than all the other planets put together; because when it is nearest to us it turns its brightest side toward us, and at all times it is comparatively free from heavy clouds, so that it may be readily observed. Mars, too, may quite frequently be seen with the naked eye. It comes into opposition with the Sun—that is, it rises in the east as the Sun sets in the west, and *vice versa,* showing its brightest and best at midnight, —about once in two years. When it is nearest the Sun—at *perihelion,* as astronomers say,—it shines with a bright, steady, untwinkling red light. And from its color comes its name: red is the insignia of war, and Mars was the war god of the ancient Greeks.

Mars is only a little over half as large as the earth; its diameter is 4,230 miles; its day is seventeen minutes longer than our day, and its year is 687 of our days. If the Martian year were the same length as ours, Mars would come into opposition with the Earth at the same time every odd year. But the difference of forty-three days less than two of our years causes this opposition to vary about one month from one time to the next. By reference to a good almanac you will learn exactly in what month to expect Mars. Should the favored time be August or September then the planet will be at perihelion, and the sight will be very brilliant; but if the opposition is scheduled for February, the planet will be at *aphelion*— the farthest from the Sun— and will shine with lessened brightness.

Strangely enough, the red color which distinguishes Mars is more noticeable when seen by the naked eye than when viewed through the telescope. The instrument shows the red of the planet splotched here and there with a dark greenish color. Astronomers think that the reddish parts are continents, which are believed to be largely desert land, while the green parts betoken seas. If this be true, then Mars has vastly more land than water,—just the reverse of the conditions on the Earth. Another interesting feature is that the Martians, if there are any such people, can go from water to land without crossing wide oceans as we must do. The surface of Mars is relatively flat, and it is thought that for the most part the waters are not very deep. Extensive regions which show shades of orange and brown are thought to be marshes. Undoubtedly Mars is an older planet than the Earth, and astronomers agree that its areas of permanent water are gradually diminishing. White spots at the polar ends of the planet are thought to be ice caps, because they

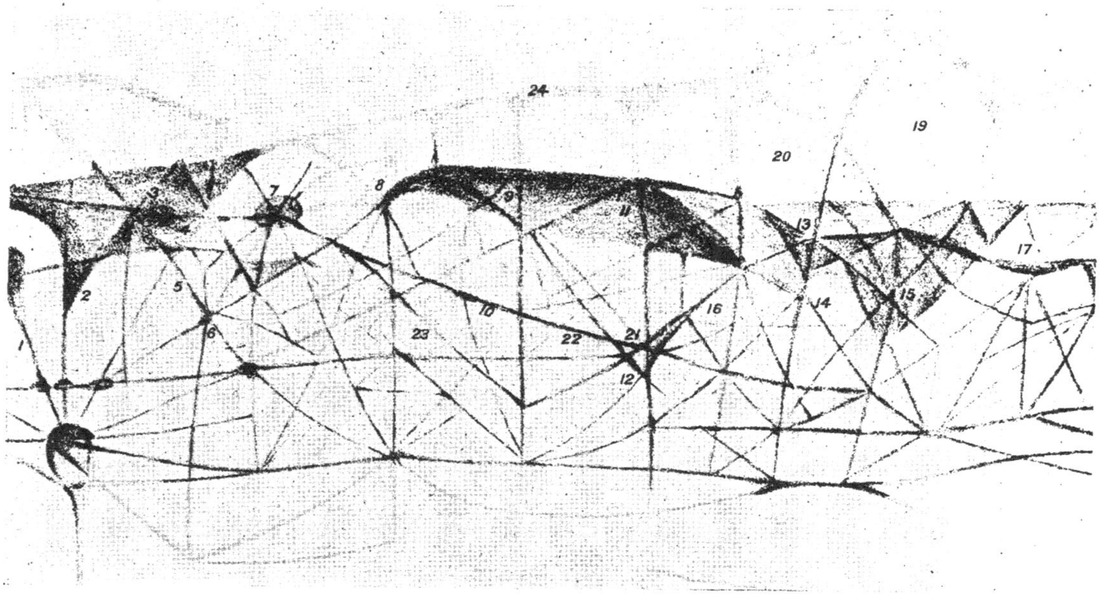

MAP OF MARS, 1896-97

1. Fastigium Aryn.
2. Margaritifer Sinus.
3. Mare Erythraeum.
4. Aurorae Sinus.
5. Ganges (Canal).
6. Lunae lacus.
7. Solis lacus.
8. Sirenius lacus.
9. Mare Sirenum.
10. Eumenides (Canal).
11. Mare Cimmerium.
12. Charontis lacus.
13. Mare Tyrrhenium.
14. Syrtis Minor.
15. Syrtis Major.
16. Cerberus (Canal).
17. Mare Icarium.
18. Edom promontorium.
19. Hellas.
20. Ausonia.
21. Trivium Charontis.
22. Orcus (Canal).
23. Pyriphlegethon.
24. Mare Chronium.

always increase in size in the Martian winter, and fade when the Sun comes again, just as in our own polar circles. Indeed, Mars shows materially the same seasonal phases as the Earth.

There has been endless speculation as to whether or not Mars is inhabited, and the majority of astronomers and scientists think that it is. But it is probable that the Martians are a people constituted differently from ourselves. Indeed, they needs must be; for the atmosphere of Mars is much thinner and rarer than our own; besides, at sunset, the temperature falls alarmingly, reaching, at times 100 degrees below zero. Edmund Perrier, director of the museum of the Jardin des Plantes, in Paris, gave to the world the first picture of the Martians, as he conceived them. "The men on Mars are tall because the force of gravity is slight," he said. "They are blond because the daylight is less intense. They have less powerful limbs. Their large blue eyes, their strong noses, their large ears, constitute a type of beauty which we doubtless would not appreciate except as suggesting superhuman intelligence." That the Martians, being an older race, are beyond us in intellectuality may well be believed.

Much excitement was aroused in the scientific world, in 1877, when Schiaparelli, the Italian astronomer, announced his wonderful discovery of streaks, or *canals* passing here and yonder on Mars. Many astronomers have since verified the truth of this statement, among the most enthusiastic being the late Professor Lowell, director of the observatory at Flagstaff, Arizona, who looked upon the canals as direct evidence that Mars is inhabited by a people who thoroughly understood all the principles of intensive irrigation. To him it was plain that the temperate regions of Mars were an arid desert, and that to sustain their life the Martians had been compelled to intersect their country with a system of canals, and to bring water from the polar ice-caps. In 1914, the professor's belief was further strengthened by the discovery that two new canals had recently been built. "We have actually seen them formed under our eyes," he announced, "and the importance of it can hardly be overestimated. The phenomenon transcends any natural law, and is only explicable so far as can be seen by the presence out yonder of animate will."

All astronomers, however, do not agree to this interpretation of the black streaks which criss-cross the planet. Nor can their true explanation be verified until more powerful telescopes are made, or until that happy day when wireless is established with Mars. "To talk to Mars," says Nikola Tesla, "seems to me only a matter of electric power and perseverance." Professor Albert Einstein, whose relativity theories have lately made such a stir in scientific circles, thinks that the best medium for communication with the planets is the light ray. Photography, too, may develop further interesting features. Perchance some one of the boys or girls who read these lines may later solve this question of life in other worlds than ours. Meantime, even the best astronomers tell us that, after all, the merest amateur in astronomy really knows as much about it as they do, which is nothing at all.

Early astronomers thought Mars an unattended body, like Mercury and Venus; but, in 1877, two satellites were discovered. These moons of Mars are about seven miles in diameter. They are the smallest bodies known in the solar system, with the exception of the asteroid Eros. Professor Newcomb compares their size, as we see them, to a small apple viewed above Boston from a telescope in New York City. They can be seen only when the planet is near opposition, and then only through the large telescopes. Because of the brightness of the planet, we are told that the outer moon is easier to see than the inner one, although the latter is really the brighter of the two. Phobos and Deimos these two satellites are styled, in honor of the sons of the old war god. Phobos, the inner moon, is 4,000 miles from the planet, and goes around it in seven hours and thirty-nine minutes. Phobos, therefore, must seem to the Martians to rise in the west instead of in the east, as our Moon does. Furthermore it crosses the heavens three times in one day. It gallops across the sky, overtakes Deimos and eclipses him, we are told, and runs through all its phases in eleven hours. What a strange world where the moon could be seen early in the evening at the first quarter, and three hours later at the full! Deimos takes a much more leisurely course than his brother; in fact he is about two and one-half days making his journey. It would certainly be worth a journey to Mars just to see these wonderful moons! It is doubtful, however, if either of these orbs furnishes enough light to illuminate the evening skies. Perhaps we would do just as well to observe them from here, apparently playing hide-and-seek with their planet, peeping out now on one side, now on the other.

One astronomer, writing of the view of the Universe from the red planet, says: "Jupiter is magnificent from Mars; he appears to the Martians half as large again as he seems to us, and his satellites should be easily visible to the naked eye. Saturn is likewise very brilliant. Uranus is easily visible, and they might have discovered Neptune before we did. They must have distinguished with the naked eye a large number of the small planets which revolve between their orbit and that of Jupiter. Mercury, drawn closer to the Sun, and lost in his rays, is almost impossible to distinguish. Venus appears to them as Mercury does to us.

As for the Earth, how do we see it? . . . We are for that planet [Mars] a brilliant star presenting an aspect similar to that which Venus presents to us, preceding the dawn, and following the twilight; in a word, we are to the inhabitants of Mars the shepherd's star." [1]

THE ASTEROIDS, OR MINOR PLANETS

The asteroids, or planetoids, as they are sometimes called, form the division between the inner and the outer planets, and the story of their discovery is another proof of the value of good guessing in the astronomical field. Such a vast space exists between Mars and Jupiter that it was early felt that a planet should be found revolving between the two. For more than 300 years the search was carried on, and astronomers had about lost hope of finding the missing one, when Bode, a German astronomer, put fresh stimulus into the proceeding by a set of figures showing a curious relationship in planetary distances. By his rule—known as Bode's law—there was a missing link just where astron-

1 "The Romance of Astronomy,"—Macpherson.

omers had long reasoned a planet should be found.

Accordingly the Zodiacal region of the heavens, which is the path of the planets, was divided into convenient sections for observation, and the world of astronomers set themselves each to "policing" his own territory in the sky. But there were no results. What had become of the missing planet?

Piazzi, the director of the observatory at Palermo, Sicily, found a hint of the solution, on the first night of the nineteenth century. While making observations for the purpose of putting out a star catalogue, he noted an eighth magnitude star which was new to him. Observing this star a second night, how was he surprised to note that it was in motion! It was, then, not a star; no doubt it was a comet. But it was tailless, and it had none of the comet's eccentricities. Could it possibly be the missing planet? Piazzi made a chart as best he could, and sent his information to Bode. On account of illness, the director was then forced to leave his post for six weeks, and ere his return the planet had disappeared. All feared that it had gone beyond recall, but happily mathematicians managed to calculate an orbit from Piazzi's notes, and on the last day of the century the planet hove into sight again, just as the mathematicians had figured that it would do! Ceres, it was called, in honor of the guardian goddess of Sicily.

To be sure, Ceres was only a small planet—480 miles in diameter; nothing at all like what they had expected to find. But she filled the gap, and astronomers felt that they must be satisfied. Imagine, then, their surprise, as the most of them went about other business, when three months later, Olbers, a German astronomer, discovered another planet not far from Ceres. Pallas, they called it, and it was soon seen that both of the new planets revolved about the sun at nearly the same distance.

What did this mean? For now the symmetry of the solar plan was broken. Undoubtedly the two bodies were related, and presently it was decided that at some period a planet, similar perhaps to Mercury, had existed between Mars and Jupiter. Age, or some catastrophe, had dissolved it, and the two small planets were the "remnants." This opinion was further strengthened by the discovery of Juno, in 1804, and Vesta, in 1807. No further planets having been discovered after a period of careful search, the theory was regarded as fully established, and for forty years no one thought of questioning it.

Then, rapidly one after another, due no doubt to the increased power of the telescope, astronomers here and there began to discern other small planets, and again the search for planetoids was begun with feverish eagerness. Presently so many had been definitely located as to make the effort of naming them absurd. It was suggested that the letters of the alphabet be applied, but these soon proved altogether inadequate, and the process was begun over, designating the additions *AA, BB,* etc. This, of course, was confusing, and not sufficient either, and, after a time, the sensible plan of numbering the planets was adopted. To-day the latest additions are catalogued up in the nine hundreds, and astronomers are no longer specially interested in the planetoids. Of them all, a little asteroid, called Eros, which was discovered in 1898, is the only one that has served any particular purpose. Unlike its kindred, Eros does not travel in the field between Mars and Jupiter; its orbit is so elliptical that once in thirty years it comes within the orbit of Mars, and is then our nearest planet neighbor, only 13,500,000 miles away. At this distance, it has been photographed many times, and by measurements thus secured, astronomers, who never like to rely on any one method, have been furnished with another means of calculating the solar dimensions.

It is hardly probable that any living beings are to be found on the asteroids. Certainly they would have to be of a totally different nature from any creatures known on our Earth. Suppose the planets to be of the same density as our sphere, on a planet one hundred miles wide,—one-eightieth of the diameter of the Earth— every weight would be reduced to an eightieth part of what it is here. The average man, then, would only weigh about a couple of pounds, and as Professor Ball points out, such a creature might easily possess a pair of wings and be able to fly, provided the atmosphere were conditioned like our own.

This writer gives an entertaining picture of a

game of lawn tennis under such conditions—which at best would be a difficult pastime.

"The very slightest blow of the racket would drive a ball a prodigious distance before it could touch the ground; indeed, unless the courts were about half a mile long, it would be impossible to serve any ball that was not a fault. Nor is there any great exertion necessary for playing lawn tennis on Flora, even though the courts are several hundred acres in extent. As a young lady ran to met the ball and return it, each of her steps might cover a hundred yards or so without extra effort; and should she have the misfortune to get a fall, her descent to the ground would be as gentle as if she was seeking repose on a bed of the softest swan's-down."

We know practically nothing of the real condition of the asteroids. The majority of them can be seen only through the most powerful telescopes. No traces of atmosphere round them has been found. Nor has it been possible to determine their rotation or surface conditions. Some show such variations of light that it is probable they have rugged outlines on which the light reflected from the Sun varies according to their position. We do know, however, that the old theory of their origin from a wrecked planet is without foundation. Astronomers are now satisfied that rings were thrown off from the nebula in space, and that these should have united to form a large planet. But owing to the immense attraction of Jupiter such a condensation of rings was impossible. Thus these innumerable bodies, each pursuing its own appointed path around the Sun, serve as one of the wonderful niceties of the harmony and balance existing in our Universe.

JUPITER, THE GIANT PLANET

Jupiter is the mightiest of the planets, so great indeed that all the other planets and their satellites rolled into one could not fill the space he occupies. Beside him, our Earth seems insignificant; 1,200 globes of our sphere's size would not quite cover this immense planet. Jupiter is 483,000,000 miles from the Sun. His diameter is 87,000 miles. His year is almost twelve of our years long. But his day is only ten hours in length, which means that Jupiter in spite of his immensity, is rotating on his axis so quickly that every five hours the side of the planet which is toward us presents an altogether new view.

When Jupiter is in opposition to us he presents a sublime spectacle. This occurs about once in every thirteen months, and he reaches the meridian at midnight. Just as the Sun is sinking, this big bright star rises in the opposite horizon, showing clearly at twilight and growing more brilliant as the night advances and he ascends the sky. It is impossible to mistake him. He shines supreme—a bright, particular king.

Since Galileo, in 1610, pointed his newly invented telescope—the first instrument of the kind the world had known—at Jupiter, this silvery body has been the delight of astronomers, from the merest amateur to the most expert. Even the smallest instrument will show that the wonderful radiance comes from a round disc-like body, which is obviously a planet, and will give a glimpse of the four largest satellites—moons which run through all sorts of brisk and lively changes and eclipses, and which it is thought, may possibly be inhabited, the larger exceeding the size of the planet Mercury. These moons of Jupiter, too, are interesting from another point: it was their eclipses which first suggested to astronomers and mathematicians that light had a measureable speed. To-day we know that light rushes along so swiftly that it could go more than seven times around the earth in a single second, its velocity being, in truth, 186,300 miles per second. Altogether Jupiter has nine moons. The innermost one flies around him in about eighteen hours; but the outside one takes nearly seventy days to perform the revolution.

Jupiter is of vast interest because it is evidently a world in the making. Great clouds rise from it in immense volume, showing that it is still terrifically hot, no doubt yet in a semi-liquid and gaseous condition. It is thought, however, that the light which it casts is not from its own heat. It has cooled sufficiently to reflect the Sun's rays, but not enough to maintain water on its surface. Doubtless a great part of Jupiter's size is due to its inflated, gaseous condition; for *by means of its moons,* incredible as it may seem to us, mathematicians have succeeded in weighing the great planet's mass.

They tell us that it weighs only 300 times as much as the earth.

When viewed through the telescope the vastness and diversity of Jupiter's landscape become more apparent. The equator is partly hidden by huge belts of reddish clouds some of them as big as our whole Earth. The equatorial belt is yellowish or tan color where visible, but constant cloud masses break it up. Toward the temperate and frigid zones these clouds become steel-blue and gray, showing that the Sun exercises less and less effect toward the poles.

As for Jupiter's position in reference to our universe, we are told that if it were possible for us to be transferred to the great planet, and if we could see through the cloud belts, the Earth would be scarcely visible, being only seen with difficulty in the vicinity of the Sun. The constellations in the heavens, however, would be those with which we have long been familiar; for the 390,000,000 miles which separate us from Jupiter is a mere nothing in the celestial perspective. Unquestionably the most curious spectacle from the Jovean realm would be its own marvelous moons. To these Jupiter is himself a miniature sun, giving out a certain amount of heat as well as reflected light.

Galileo's discovery of the four large moons of Jupiter was the best possible proof of the Copernican theory. Here on a plane easily discernible was a reproduction of the solar system which should convince the most skeptical. But certain prejudiced opponents of the system, it is said, refused to look through the telescope; others, and perhaps the most stiff-necked, looked and refused to credit the spectacle, saying that the satellites were in the telescope, not in the sky! The four moons, named in order of size, are Ganymede (III), Callisto (IV), Io (I), and Europa (II); but they are most commonly known by their numerals, which indicate their nearness to Jupiter.

SATURN, URANUS AND NEPTUNE

Saturn is the farthest planet that can be easily observed with the naked eye, and then he appears only as a star of the first magnitude, about the brightness of Capella in the constellation of Auriga. There is one advantage, however, when he rises at sunset he can be seen during the whole night, so that one has not only plenty of time to find him but to observe him as well. Moreover, he recurs night after night for several months. The ancients thought little of him; indeed they considered him an unlucky planet, slow of motion, dull of light, and altogether a symbol of gravity and gloom. Persons so unfortunate as to be born under his sign were sure to be dull and morose in nature, "saturnine." Under the powerful telescopes of to-day, however, Saturn takes on an altogether different hue. So far from being the least interesting of the planets, he is indeed altogether the most fascinating and unique. To the best of man's knowledge there is none other like him in our universe.

What we see through the powerful lenses is a glorious dull yellow orb, striped with belts similar to those of Jupiter, but fainter owing to its great distance. (Saturn is sometimes nearly a thousand million miles from us.) Around the planet is a wonderful system of three rings—two bright, and one of semi-transparency, called the *crepe* ring. The rings, it has been found, are made up of millions of little pieces, tiny moons or satellites in themselves. The moons may be separated from one another thousands of miles, but at this distance they show as a complete circle or ring. Moreover these magnificent rings, taken together as a plate or rim, are not less than 176,000 miles in diameter, 30,000 miles in width, and from fifty to one hundred miles thick. Our Earth could revolve on this "celestial deck," as Flammarion puts it, "like a ball rolling along a road." Moreover, the planet in the center is more than 900 times the size of our sphere. The ring system however, may be a transient feature, another satellite in process of formation.

Saturn has ten known moons, each revolving in its well-appointed orbit, and depending on him not only for heat but as the giver of light. Should any of these satellites have inhabitants, they probably regard their planet much as we do our Sun. Titan, the largest of these worlds, equals the planet Mercury in size. Themis, the smallest yet known, is discernible only by photography. Phoebe, the most distant of Saturn's satellites, is the most remarkable little moon in existence, possessing an independent spirit that is both staunch and unique. Usually members of the solar system travel *eastward,* or from the west to the east; but not so Phoebe. For some reason unknown she pursues a contrary course, going *westward,* or from east to west. One may imagine the confusion of Saturn's evening skies, with nine moons flying along, one after another, each one taking its own time to complete the revolution—the nearest, little Mimas, being over two days on the job, for Saturn's day, you remember, is only ten hours long—then contrary little Phoebe sailing along majestically in the opposite direction and openly voicing her defiance to regulation by leisurely taking *sixteen* months to perform her voyage!

Not only is Saturn remarkable for the striking spectacle of his rings and moons, but he is the lightest of the large planets, being only ninety times heavier than the earth, notwithstanding the fact that he is over 700 times larger. Indeed, says one authority, "Saturn is only equal in weight to a globe of walnut wood of the same size. In fact, if we could imagine a great ocean large enough to hold the various planets, and if we could imagine the planets thrown one by one into that ocean, Saturn would actually float while all the others would sink." Of course this extraordinary lightness of the planet can be explained in only one way; Saturn, like Jupiter, is in an early stage of development. The vast clouds which rise from its surface betoken a state of chaotic existence—violent eruptions, seething boiling masses of inflammable elements. clouds of gas, and a sweltering heat.

The planet Uranus can sometimes be seen in the spring and summer months, shining with a pale green light, if you have a good eye and know just where to look for him. It is probable that the world marked him as a star for centuries. Any real knowledge that has been had of him has, of course, come through the most powerful telescopes; though Herschel, the German astronomer who first sighted the planet, was a mere amateur, a musician by trade, who worked with a home-made telescope. Uranus is 1,780 millions of miles from the Sun, and has a diameter of 32,000 miles. His day is thought to be about eleven hours long, and his year is eighty-four of our years long. By the aid of his four satellites, Uranus, though such a vast distance from the Earth, has been weighed in the mathematical scales and found to be about fifteen times heavier than the earth. The spectroscope has proven that the planet has an extremely dense atmosphere. The two inner satellites of Uranus are about 500 miles in diameter; the outer ones are nearly twice as large. Like Phoebe, they choose to travel in contrary motion, that is from east to west, or as astronomers say their motion is *retrograde.*

While astronomers were busy studying Uranus, in order to discover as much about its condition as possible, careful observers soon took note of its eccentricity. It did not follow the exact path which the mathematicians figured that it should. Evidently something was interfering with its motion. What could it be? Was there yet another planet beyond this one? Two young students, one at Cambridge, England, named Adams, and the other Le Verrier, a Frenchman, set out to figure an orbit such as it was probable the unknown planet, if there were one, was traveling. Neither knew of the other's plans, and it chanced that Adams finished his calculations first and sent them in to the Royal Astronomer, Airy, a former professor of Cambridge. It happened that the great man was extra busy, and he put the memoranda in a pigeon-hole and forgot all about them. When Le Verrier finished his drawings and conclusions, they chanced to meet Airy's eye. Happily the professor now remembered young Adams' offering, and on bringing it to light, found that the two young men had arrived at the same conclusion. This was a marvelous coincidence; probably nothing more. But the professor made haste to have their suggestion tested, and there indeed was the disturber of Uranus, a planet showing a greenish disk of light. About the same time, a like test was made at the Berlin observatory, by Dr. Galle, at Le Verrier's request with the same results. There could be no doubt, then, that a new addition had been made to the solar system, and the new wanderer was shortly christened Neptune.

The planet is altogether invisible to the naked eye, but it is sufficiently bright to have been occasionally recorded in old star maps as a star. Indeed, some fifty years before it was named a planet, the astronomer Lelande included it in a list of stars he was making, setting down in his diary some observations which later marked him as a sorry blunderer. It seems that he had observed the star and noted its position on May 8, 1785. Two days later he noted that the star was not in the position he had indicated. He, therefore, concluded he had made a mistake the first time, and accordingly moved the star on his plot. But the point of light was not a star; it was a planet. Its movement should have revealed the fact to Lelande, but he unwittingly let this proof slip by, and thereby catalogued himself in the world's records as a heedless astronomer, instead of a man of renown. Seldom has anyone more narrowly escaped fame!

Neptune is at such a remote distance,—half as far again from the Earth as Uranus—that we can scarcely hope ever to know very much about him. His diameter is 36,000 miles; he is, therefore, a little over four times the width of the Earth, thus holding the same proportion to our sphere as the Earth does to the Moon. Like his great kindred, Neptune is hidden in dense clouds, and it has been determined that his weight is only one-fifth of what it would be if he were made of earthly elements. He is, in all probability, comparatively in his infancy. His year is known to be 165 of our years long, but the length of his day has not been determined.

Are there other planets outside of Neptune? Can we consider it, as Macpherson queries, "the frontier of the Sun's domain"? Astronomers are not yet ready to answer this question. Because of unexplained influences affecting certain comets, many believe that there is still another planet in the Sun's family. We have only known of the existence of Neptune for about seventy years. Perchance ere the century ends some one may be ready with a solution of the query.

V

THE MOON, A DAUGHTER OF THE EARTH

THE ancients named the first day of the week Sun-day, in honor of the sun-god. Likewise, the second day was named after the next most important celestial body—the moon—Moon-day, which common usage has shortened to Monday. Though the Moon is our nearest neighbor in space, so near that its distance may be counted in thousands of miles, it is yet a long way off—about 240,000 miles. "If you were to wrap a thread ten times round the equator of the Earth," says Professor Ball, "It would be long enough to stretch from the Earth to the Moon. Or suppose a cannon could be made sufficiently strong to be fired with a report loud enough to be audible 240,000 miles away. The sound would only be heard at that distance a fortnight after the discharge had taken place." An express train, traveling at the rate of sixty miles an hour, and stopping neither night nor day, would reach the Moon in about five and one-half months.

No explorer has ever visited our satellite, and it is not probable that one ever will; yet we really know more about the geography of the Moon than we do of some parts of the Earth. Take the dark continent of Africa. Here are wide tracts of forests, great lakes, and lofty mountains of which little is known. While, on the side of the Moon nearest us, there is scarcely a spot as large as the average county which has not been mapped and photographed many times. Astronomers have painstakingly built up great charts of the Moon, as large as good-sized garden plots. One of the most notable of these, the work of Johann Schmidt, was begun when the great German astronomer was only a lad of fourteen, and was completed shortly before his death, forty years later. All points of lunar interest are, of course, carefully named on the various maps and charts, and astronomers are probably more familiar with the terms Eratosthenes and Tycho than you are with Vesuvius and Kilauea.

A favorite theory is that the Moon was once

a part of the Earth. To throw off such a mighty part of its bulk—one-fourth of its size, to be exact—the Earth must have been whirling on its axis at a terrific rate; making a complete revolution about once in two hours, it is figured, instead of once in twenty-four hours, as it now does. Thus the centrifugal force was enough to overcome the force of gravitation, and the Moon was torn out from the Earth's side and thrown into an orbit of its own. Certain astronomers believe that the "hole" thus left in the Earth is now filled by the waters of the Pacific Ocean, and there are some very likely proofs: (1) If the waters of the Pacific could be rolled into a ball, their bulk would be just about the size of the Moon. (2) The coasts which surround this ocean are roughly circular in form. (3) There is a great similarity between the volcanoes of the moon, and those of California, the Hawaiian Islands, and Japan; but what caused these volcanoes can only be guessed at. One theory is, that when the separation took place, great showers of melted lava were thrown out and fell, settling in gigantic masses both on the Moon and on the Earth in the vicinity where the pent-up gases issued. Many of the volcanoes so established are still active on the Earth, but those of the Moon are dead, because the Moon being so much smaller than the Earth cooled more rapidly, and thus the internal fires necessary for volcanic outbreaks were soon quenched.

Though the Moon is the Earth's daughter, she is not a world like ours is to-day. She is a type of what the Earth may be ages hence. For the Moon is cold and dead; her life fires have long since burned out. She is a "closed chapter in the book of time." As Flammarion points out: "In space there are both cradles and tombs." Jupiter, Uranus, and Neptune, as we have seen, are in the cradle stage. The Moon is one of the tombs of the Universe. Bleak and bare and still she is; without air, or water, or life of any form. Besides her mountains and craters and consequent valleys, the Moon has great gray stretches, which the early astronomers thought to be seas and named them accordingly, the "Sea of Serenity," "Sea of Tranquillity," etc. It is now known that there are no seas on the Moon. The gray stretches are simply vast dead planes; perchance ocean beds from which the water has long since disappeared. But the names have been retained as a matter of convenience. The mountains of the Moon have been called after those of the Old World. There are the Alps, the Apennines, and the Carpathians. The two highest ranges, Doerfel and Leibnitz, are from 30,000 to 36,000 feet, much exceeding the highest mountains on the Earth. More than forty lunar peaks have been noted which exceed Mont Blanc. Todd explains the greater height of the mountains on the moon as due to the lesser surface gravity of our satellite. Perhaps the reader may wonder how the height of mountains on the moon can be determined? The answer is by measuring the shadows which they cast.

The craters on the Moon get their names from ancient scientists: Tycho, Ptolemy, Copernicus, Plato, Archimedes, etc. Some of these craters are enormous. The walls of Copernicus, for example, are about fifty miles thick, and over two miles high. All about this great disc the surface is rugged in the extreme, and from its center rises a cluster of conical mountains.

Triesnecker, another remarkable crater, has a mysterious system of cliffs or chasms, a mile across at their widest point, and some 300 miles in breadth. Some astronomers credit this wondrous spectacle as the path of an ancient river bed. Similar to the craters in formation are the great walled plains, often 150 miles across, surrounded by mountainous ramparts, rising often 12,000 feet above the enclosed plain. *Rills* is the name given to a series of great yawning chasms, which run for miles, in fairly straight lines, over craters, mountains and plains, across portions of the surface of the Moon. Some astronomers regard the marvelous array of irregularities on the Moon's surface as proof that neither water nor atmosphere has ever been present on our satellite.

The Moon takes as much time to turn on its own axis as it does to go around its orbit—about twenty-eight days. So that, like Mercury and Venus, the Moon's day and year are of the same length. For this reason, too, the same side of the Moon is always turned toward the Earth; and thus it follows that the lunar days and nights are about as long as fourteen of our days.

The Moon's actual distance around the Earth is a little longer than its distance around its orbit from a certain fixed star back to the same star again; thus the Moon, like the Earth, has two days, or rather periods, as the astronomers term them: the *synodic period* and the *sidereal period.* The synodic period is equal to about twenty-nine and one-half of our days, and from this division of time the ancients got the idea for the month.

Since the same side of the Earth is always turned towards the Moon, of course it follows that we really know only one-half of the Moon. But as she sways slightly in her orbit, astronomers occasionally catch glimpses of the other side, and in this manner it is judged that the unseen part is in keeping with the side which is in full view. The Moon is not always the same shape, nor does it always appear at the same spot in the sky. The "Inconstant Moon" Shakespeare terms her. The changes in the Moon's shape are called its phases, and these phases are governed by the Moon's position with regard to the Sun. For the Moon is herself a dark body, and the light she shows is all reflected from the Sun.

A simple little experiment will show you how it is that the Moon shows phases, for of course she does not change her shape; the whole round Moon is always there, only part of it is in the shadow. Hang a rubber ball from the ceiling in the center of a dark room. Then set a lamp to represent the Sun in such a manner that its beams fall squarely on the side of the ball, which you are to fancy as "the Moon." Now, as you look at the Moon from squarely in front of it, you see the whole bright side. It is full moon. Moving a little farther to one side, either way, the Moon looks nearly full—foot-ball shape, in fact. This is the *gibbous moon.* Going a little farther, we see only a quarter of the ball. This is the half moon, or as it is more commonly called the *first* or *last* quarter, depending on whether it is west or east of the Sun. A little farther and only a slim crescent is seen. This is the *new* crescent, or the *old* crescent Moon, depending on direction as above. Going on still farther you find the Moon between you and the Sun. Its dark side is toward you, and it is therefore invisible. This is the new moon.

You will find it most interesting to follow a

course of the Moon's phases. Watch your calendar for the date when the new moon is due. You won't be able to see it, of course, but two or three days later you may expect the new crescent low in the western sky, with her horns, or cusps, turned toward the east. Now, too, you may perhaps catch a glimpse of the "old Moon in the new Moon's arms." This is an interesting phenomenon. What we see is the dark globe of the Moon apparently filling the slender silvery crescent-like arms of the new Moon. But probably you would never guess why this is to be seen. It is *earth-shine.* The Suns' rays are reflected from the Earth to the Moon, so that the Moon is lighted with earth-shine just as the Earth is lighted with moon-shine, and it is these reflected rays that light up the dark part of the Moon, causing it to appear embraced by the crescent made by the Sun. Think what a marvelous journey these light rays have had! From the Sun to the Earth, then back to the Moon, and down to the Earth again!

The Moon moves eastward among the stars of the zodiac at the rate of about thirteen degrees each day; hence she appears to us about fifty-five minutes later each night. When the Moon is near the quarter phase, her shape is a good guide to mark out the path of the ecliptic. Join her horns by an imaginary line; then a line standing perpendicular to this line, extended both ways, will very nearly point the path followed by the Sun. The small crescent Moon shines through only a small part of the night, the half Moon gives its light for half of the night, and the full Moon, although it rises a little later each night, shines all night long. About the time of the autumnal equinox, September 21, when the days and nights are equal, the full Moon rises at nearly the same hour for several nights in succession. This is called the *Harvest Moon.* The October full Moon is the *Hunter's Moon.*

All of you have seen "the Man in the Moon," and doubtless you know, too, that the markings which seem to portray the human face are, in truth, actual configurations of the Moon's surface. The ancients were much exercised about these dark markings. They thought at first that the Moon merely served as a huge mirror in which to reflect the rough contours of the Earth. Indeed, even now, those who know little about astronomy often fancy they see the outlines of Europe, Asia, and Africa on the bright face of the moon. One may find out many things about our satellite just by watching it with the naked eye. One of the best tests of the eyesight is to be able to detect the great plain called *Grimaldi.* It is a dark oval spot, containing some 14,000 square miles, so astronomers say, and there are huge mountains flanking it on the sides. By the aid of a small glass, the field of the Moon is made much more interesting. One of the most fascinating phenomena is to be seen when the Moon passes over, or *occults* a star, causing it to disappear instantaneously.

Long ago the Moon was the best means the travelers had for reckoning time at night. At full Moon the orb is due directly south at midnight. Every night before the Moon is full one must subtract fifty-five minutes; that is one night before full moon it is due south at 11:05 P. M; two nights before full moon it will be due south at 10:10 P. M. After the Moon is full it will be due south fifty-five minutes later, that is one day after full moon the orb is due south at 12:55, and so on. To-day the Moon is of such immense importance to navigators that many astronomers are needed to keep accurate record of her movements. It is the attraction of the Moon for the Earth that produces the tides. The hours for "high" and "low" tides are carefully computed by the astronomers and form a very necessary part of the nautical calendars. The Sun has some little influence on the tides, too, and when the Sun and the Moon are in line, as at full and new Moon, the tides are highest. We call these *spring tides.* At half-moon, the Sun and the Moon are exercising a pull in opposite directions. Then the tides are the lowest. Such tides are called *neap tides.* The tides are Nature's scavengers of the sea coast. Without their energetic scrubbing and cleansing, rising twice in every day and night, our beaches would be far from the spots of delightful freshness that they now are.

Suppose that we might be able to hail a passing comet and hie ourselves off to the Moon: what would we really find there? We should have to carry air with us, or we could not possibly live even to reach the Moon, much less get our breath after we got there. At the height of three or four miles above the Earth, the atmosphere grows very thin, and at ten miles up, unless we had life-giving oxygen with us at command, we should die. It might also be a good plan to have water with us. But we won't bother with these essential details. Just suppose the problem solved and ourselves landed on the Moon. Will we dare set foot on the rocks? No doubt they are unspeakably cold, all about is freezing. There is a possibility that we may turn immediately to pillars of ice! There is no air-blanket to hold the heat from the Sun, so that luminary is of no possible use, so far as heat is concerned. Astronomers have determined that during the long lunar nights the temperature must drop to 300 degrees or more below zero. We shall certainly stay but a moment!

How extraordinarily light we feel! And with reason: for objects on the Moon weigh only one-sixth what they do on Earth. We scarcely feel the weight of our big silver watch, as we draw it out to see the time. A stone that we could not possibly lift at home is picked up without the slightest effort. And as we step hastily off to have a look around, we stop in amaze, for it is as though we had donned the Seven League Boots! One of our party motions us to try a jump; but we are not to be tempted. We feel very sure we could vault that great rock yonder without half trying. Indeed, more than likely we should land away over in Copernicus or—or Archimedes! Of course, no one makes any attempt to talk. For as there is no air here to carry the waves of sound, we could not possibly be heard. Also we are careful to stand in the direct sunlight. Indeed we hardly dare to look into the shadows. They are deeper and blacker than anything that could possibly be imagined on the Earth. We feel sure that if we should step off into one, we should be as certainly lost as if we had fallen into one of the deep pits. We should be invisible to our comrades, and whistling or shouting would not avail.

Our attention is attracted to the wide Universe. Comparatively close to us is a moon,

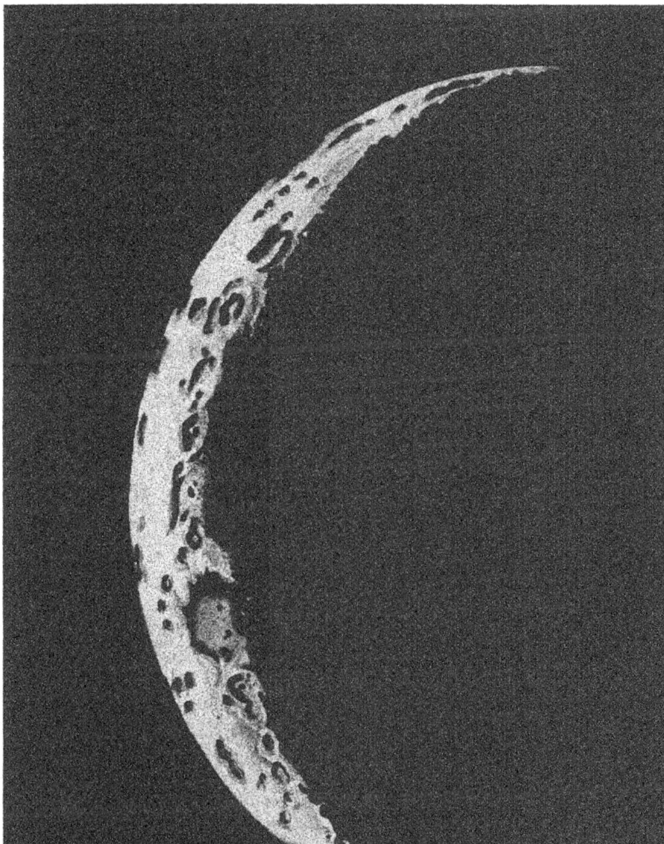

THE MOON· THIRD DAY

THE MOON—REGION OF CLAVIS AND TYCHO

about four times as broad as the satellite we are used to gazing upon. It has wonderful polar caps of ice and snow, vast oceans, mountains and plains, and great hazy cloud areas. Somehow, there is a very familiar look about its general outlines! Can it be that this wondrous moon is our earth? Yes, indeed. If we were to stay long enough, we should see it turn on its axis every twenty four hours and run through all its phases from crescent, quarter, gibbous, and full back to crescent again, just as we see our own Moon doing when we are at home. Millions of miles out in space is the Sun, and because of the absence of air, we see the great orb of day in all his wondrous magnificence,—sunspots, fiery prominences and marvelous corona, all in dazzling array. What an unspeakably gorgeous spectacle! The stars, too, seem much bigger and brighter than they do from the Earth, and there are many more of them—as many as we see at home when looking through a three-inch telescope.

But, notwithstanding all these outward splendors, it is a relief to embark from this cold, dead world, with its endless succession of jagged rocks, great gray barren plains, and extinct volcanoes—this tomb rolling onward through space.

VI

COMETS, OR THE GHOSTS OF SPACE

THE name comet comes from the Greek word *hair,* these ancient people having styled the comet as a hairy star, because of its long luminous tail streaming out like a bright veil of woman's hair. For generations the comet was everywhere looked upon as a harbinger of great tidings. Early preachers taught that it was composed of the sins of mortals, which, ascending to the sky, were set on fire by the wrath of God. Copernicus scoffed at this doctrine; Kepler claimed that the comets originated, ran their course, and died; other early astronomers, too, did their best to show that as the comets were superlunar—that is beyond the Moon—they could not be intimately concerned with war, pestilence, or famine. But the ignorant multitude gave these assurances small heed. The downfall of Nero was supposedly heralded by a comet; in the time of the Norman Conquest, a comet appeared spreading its glorious tail across the sky, a forerunner, it appeared, of the subsequent victory of the Normans. Another comet, seen in 1456, supposedly had a connection with the capture of Constantinople by the Turks. As late as the seventeenth century, an illness among the cats in Germany was ascribed to the appearance of a particularly freakish comet.

But as science became more advanced, and telescopes more numerous, it was found that for one comet seen by mortal eyes there were hundreds which could only be seen through the glass. Obviously, then, the old theory would have to be abandoned. But whence came these capricious visitors of space? It was easy to believe that they might have been drawn by the attraction of the sun. But what kept them from tumbling into the shining orb, and whither did they go? It was Newton who first declared that comets moved in orbits, just as the planets do, and subsequently a great deal was discovered about these erratic messengers of space.

To begin with, it was found that a comet is

really made up of three parts: (1) a bright head or core, called a *nucleus;* (2) a hazy light layer, called a *coma;* (3) a luminous *tail,* which waxes and wanes in a most remarkable manner. Moreover, by means of the spectrum, the very substances which make up the nucleus, the active part of the comet, have been determined. Chief among these are iron, carbon, sodium,—all materials that are found in the Sun and the planets, and materials with which we are all familiar. Sodium perhaps you may not recognize; but it makes up the bulk of the salt in the sea. These materials appear in the comet in bunches of stones, which are held together as they speed through space by that wonderful agent known as attraction. As a comet draws near to the Sun its elements begin to get hot and to throw off burning vapors or gases, which make up its coma, and then stream off in luminous vapors forming its tail.

A comet without a tail is a very poor affair, and astronomers give it small heed. But one cannot always tell about comets. When first sighted heading toward the Sun, a comet has no tail worth noting. Astronomers, however, keep it covered. They know that if the comet is not a poor burned-out affair, there are likely to be marvelous effects as soon as the Sun's heat begins to be felt. If the tail does begin to develop, it shoots out from the head with enormous rapidity, and often grows to many million miles in length as the comet comes on with terrific speed. Once the messenger has made his obeisance to the Sun, however, and headed the other way the tail begins to diminish, and presently the comet vanishes into space as undecorated as when it was first heralded.

Comets have been known with two, three, and as many as six tails. The comet of 1882 whose wonders astronomers never tire of relating, had a marvelous tail 100,000,000 miles in length. Could this comet's head have been placed at the Sun, its tail would have streamed not only to the Earth, but out into space across it! Naturally one would expect a comet's tail to follow its head. It does, so long as the comet is rushing toward the Sun, but when it turns to go the other way, the tail sweeps on in advance. Thus, you see, a comet's tail always points away from the Sun. "Why?" queried the astronomers. But it was many years before the real reason was discovered. Now we know that the Sun not only attracts comets but that his light exercises a pushing or repulsive power on minute particles. Just what this repellent power is has caused much argument. Some scientists say that it is electrical; others that it may be due to what is called *light pressure.* At any rate, we know that it only affects the very smallest, most insignificant particles of matter. The planets and moons are in nowise troubled. This pressure, then, or repellent force, whatever its nature, drives back the tiny vaporous particles which make up the tail, causing it always to point away from the Sun. It is this pressure, too, which is believed to make up the seeming number of tails. As we have seen, a comet's nucleus is composed of certain elements. The vaporous particles of these elements, no doubt, have different densities. Consequently the repellent power of the Sun affects these unequally, thrusting the diverse elements out at different angles, so that one tail may be due to iron vapor, another to sodium, and so on.

It happens that there are other bodies in the sky very like comets; we call them nebulæ. So that just as a planet is liable to be mistaken for a star, so a comet is liable to be termed a nebula. It chances also that the test in the two instances is the same. A nebula always stays in the same place, like a star, while the comets and planets are ever on the move. There are several ways to distinguish a comet from a planet. In the first place, it must be remembered that the majority of comets are to be seen only through the telescope; the really brilliant comets that can be seen in a lifetime may be counted on the fingers. Again, comets never burst into view, wondrous and beautiful; no matter how great they may grow to be, they are first seen as dim little patches of light. They travel with exceeding swiftness. Indeed, by watching one steadily for a time it may be seen to slip past the stars it is first noted among. Each night, too, an approaching comet grows larger and brighter, and its tail presently begins to make its appearance. When it reaches this stage, the comet is more wonderful observed with the naked eye than through the most powerful telescope. Its head glows brighter than the brightest planet, and its tail spreads out in a glowing arc which lights up the whole heavens. The planets, you remember, are not found outside the zodiac. The comets shoot in and out of the solar system from and to every direction, in apparently the most erratic and method-mad fashion.

They are the witches of the sky, fantastic and changeable. We can almost fancy that we see the broomsticks! But staid astronomers are not at all in love with their antics. How many pages of painful figures have they wrecked! Here is an orbit and schedule carefully worked out. Miss Comet is due to make her bow at such and such a time. The fact is published. But she does not appear. Then, when least expected, along she may whisk, taking the world by surprise—and the astronomers with the rest.

The nearer a comet gets to the Sun the faster it travels, frequently rushing along more than a thousand times swifter than the speed of a rifle ball. One case is on record of a comet which had been coming in for an incalculable time toward the sun, suddenly being drawn forward at such a tremendous rate that in two hours it had speeded round the Great King and started back into the depths. Of course, these terrific outbursts do not last long. "A pace which near the sun is 20,000 times that of our express trains diminishes," we are told, "to 10,000 times, to fifty times, to ten times that pace; while in the outermost part of its path the comet seems to creep along so slowly that we might think it had been fatigued by its previous exertions."

An old astronomer calculated that there were probably more comets in the sky than fishes in the sea. To date, however, only about 1,000 comets have been registered. To be observed, it has been estimated, that a comet's head must be at least 15,000 miles in diameter. The average is from 25,000 to 100,000 miles, and the greatest comet yet recorded is that of 1811, measuring over 1,000,000 miles—fourteen times larger than Jupiter, the largest planet!

Where these strange visitors, these ghosts of the skies, come from, and the paths that they travel—for there is really order in their wild onrush—is one of the most interesting problems of astronomy. Suppose there should be at this moment, drifting thousands of miles out in space, a torpid bit of the novel material of which comets are made. The ever-reaching beams of the Sun, even at this vast distance, make themselves felt. And presently the little loiterer finds itself stirred to the depths. Lazily it rouses up and begins slowly to answer the impelling invitation. Its movement is only slightly increased yet, it may be centuries, a thousand, ten thousand years before it reaches the solar system; then its speed will rise prodigiously, growing ever faster as it draws nearer until, by and by, it has paid tribute to the Great Ruler and dashed onward in its career.

If it is the Sun that has called the comet, why does it not dash straight into his alluring beams? For one reason, because it had a motion of its own before the Sun's attraction was felt, and according to the laws of motion it is impelled to follow its own course. But surely if the attraction of the Sun is strong enough to call the comet such a vast distance, it ought to be able to hold the visitor when it arrives. Yes, but consider: the comet dashes up at a speed

20,000 times faster than the fastest express train. Moreover, the comet is several thousand miles away from the sun; the swift-traveling comet previously mentioned, which went by at perihelion at the amazing rate of 366 miles per second, passing around the Sun in two hours, made the nearest known approach to the great orb—30,000 miles. The Sun does not stop a comet, but it changes its direction.

Three forms of path are possible to comets. What these are may best be gleaned from the illustration. Only those comets which follow the elliptic path can be considered members of the solar system. The other two paths are open curves, and once the comet has swung in and paid his ghostly cometary tribute, he will never come back. There is one caution, however: Astronomers have found that these two latter curves may sometimes be in reality gigantic ellipses, and that in the course of a thousand years or more the wanderers along these paths may heave into sight again. Of course there must be some method of determining this, and presently the mathematicians proved that when a comet going

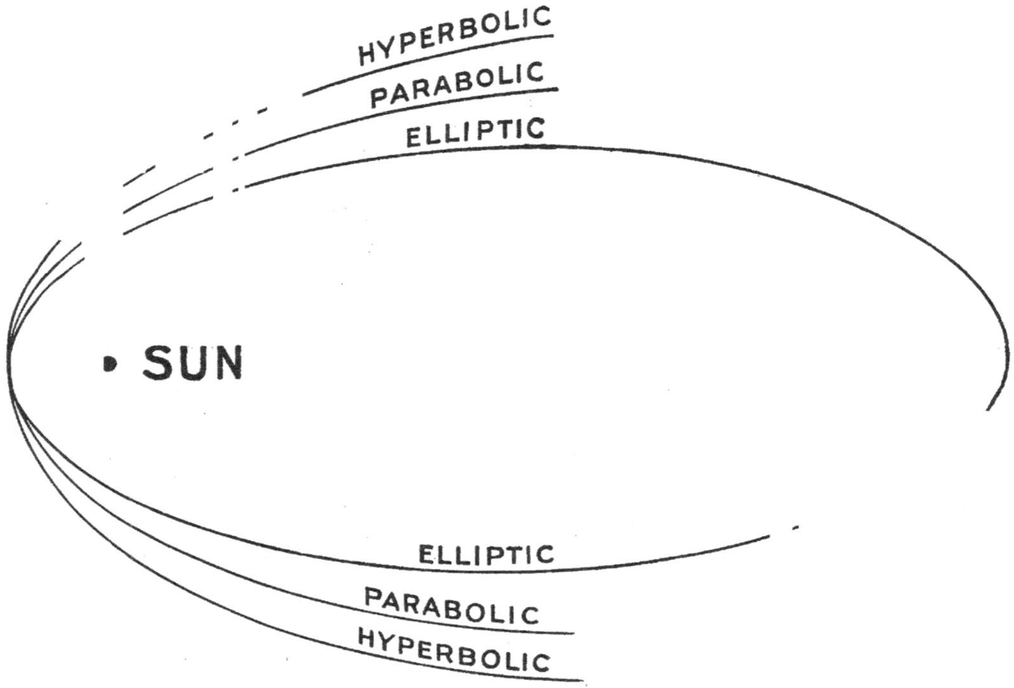

THREE FORMS OF PATH POSSIBLE TO COMETS

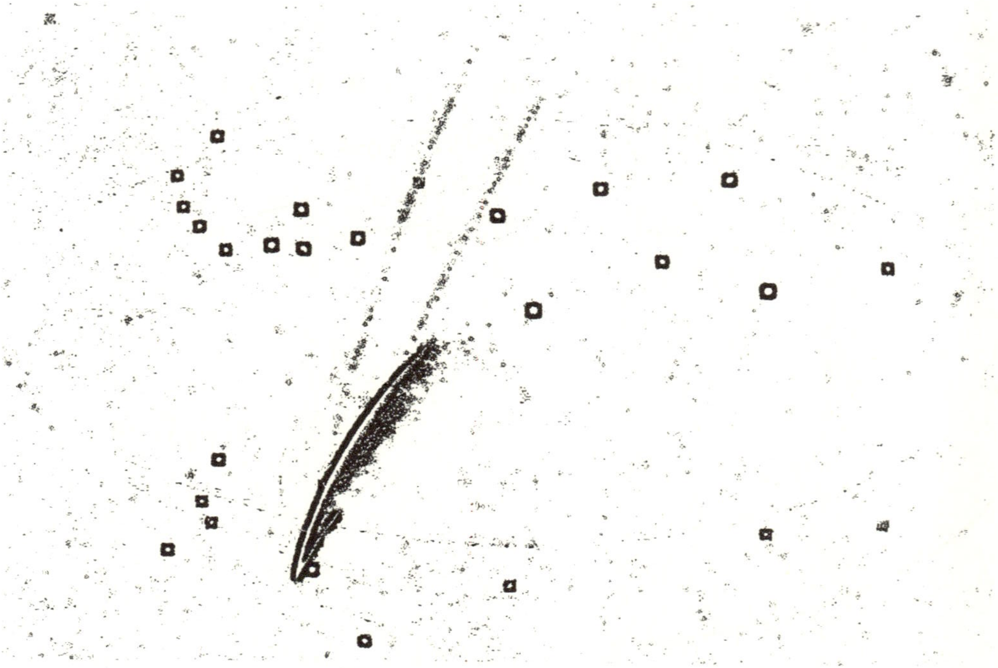

COMET OF DONATI, OCTOBER 5, 1858

at a speed exceeding twenty-six miles per second is traveling an orbit distant ninety-three million miles from the Sun it will never come back. Very often the velocity of a comet is so near this figure, that astronomers are in doubt concerning its return.

At nearly every observatory there are scouts detailed to watch for comets. Immediately one is sighted, it is at once charted, and investigations commence to see whether it is a known or an unknown visitor. And here no end of complications may arise! In the case of the planets and asteroids, certain drawings and specifications, and above all photographs, help to establish identity. But not so with the comets. Their shape and size may vary with each and every appearance. Now one will have a long straight tail; again a short fan-like appendage; again a forked tail, or none at all. The "looks" of a comet, therefore, are no aid at all in identifying it. The more certain way to do so is to trace its pathway through the heavens—and even that is not certain; for it frequently happens that during its erratic wanderings it comes in contact, or nearly so, with other heavenly bodies, and is hurled millions of miles out of its beaten path. Sometimes it loses its way altogether. Again it may be shattered into several fragments. Astronomers have frequently noted two or three such wanderers along the general path that one known comet was supposed to take. Then the question arises as to whether they are new bodies, or merely sections of the old one widely scattered.

About thirty comets in good and regular standing are now known, with periods of less than 100 years, the shortest being Encke's comet, three and one-half years, and the longest that of Halley's comet, something over seventy-five years. Nearly all of these bodies are invisible to the naked eye, and at best can only be seen for a very brief period. The outer planets all have comets which revolve in elliptical orbits about them: Jupiter 18, Saturn 2, Uranus 3, and Neptune 6. And the interesting part about this feature is that these comets are captives: once they undoubtedly described great parabolas or hyperbolas. But in rushing toward the Sun, they passed so near the planet to which they now owe allegiance that their paths were deflected, and they found themselves held forever by the overmastering attraction of the giant bodies.

Generally speaking, the great comets come within our vision once and are then never seen again. Halley's comet is an exception, and this comet is also of special interest because it was by means of it that the world first became assured that comets really traveled in orbits. Halley, the English astronomer, who first sighted this comet and charted its ellipse, found that it came very near the Sun in one part of its journey and swung out nearly to the orbit of Neptune on the other. In poring over the records to see if he had found a new comet, he noted that twice before, at intervals of around seventy-five years, messengers of space had been charted in almost this same orbit. Most carefully Halley considered the matter: he knew that, from its proximity to Jupiter and to Saturn at certain points, the path of his comet was bound to be a trifle uncertain at each revolution. And a great idea came to him: he became assured that his comet and the other two which had been recorded were one and the same. If this were true, it fully established Newton's theory that comets had orbits. But Halley was a man in his prime. He knew he would not live another seventy-five years to see his belief proven. So he published his theory, adding: "If it should return according to our predictions, about the year 1758, impartial posterity will not refuse to acknowledge that this was first discovered by an Englishman." The name the comet bears fully shows that posterity, after verifying the truth of Halley's prophecy, was only too glad to perpetuate his fame. Incidentally, as a further witness of the accuracy of mathematician's figures, three calculators set the date for the return of the comet as November 4, November 11, and November 12, 1835, respectively; the comet appeared November 15. Halley's is the comet previously mentioned as appearing in the time of the Norman conquest, 1066. It is pictured in the celebrated Bayeux tapestry. The opening lines of Tennyson's *Harold* also portray this far-famed comet:

First Courtier. "Lo! there once more—this is the seventh night!
Yon grimly-glaring, treble-brandish'd scourge of England!

Second Courtier. "Horrible!

First Courtier. "Look you, there's a star
That dances in it as mad with agony!

Third Courtier. "Ay, like a spirit in hell who skips and flies to right and left, and cannot scape the flame.

Second Courtier. "Steam'd upward from the undescendible Abysm.

First Courtier. "Or floated downward from the throne of God Almighty.

Aldwyth. "Gamel, son of Orm, what thinkest thou this means?

Gamel. "War, my dear lady!"

The question arises: Is there any danger of the Earth colliding with a comet? Since comets have a reckless habit of dashing into the solar system at any time and from any direction, it is quite conceivable that such an occurence might be possible. However, astronomers have figured that but one collision of this nature is likely to take place in a space of 15,000,000 years! "If one should shut his eyes and fire a gun at random in the air," we are told, "the chance of bringing down a bird would be better than that of a comet of any kind striking the Earth." More than once the Earth has been known to pass through the tail of a comet, and in each instance few besides the astronomers were aware of the event. In June, 1861, a brilliant comet appeared between the Earth and the Sun, about 14,000,000 miles from our planet, while its tail stretched way beyond us. All day the heavens had a yellowish tinge like that of early dawn, and the Sun shone feebly, though the sky was cloudless. At seven o'clock dusk came on, and lamps had to be lighted. Now, too, a golden-yellow disc, half-hidden in a filmy veil, appeared in the sky—evidently the Sun's rays had prevented its being seen earlier. This was supposedly the comet's head. A witness described it as "though a number of light, hazy clouds were floating around a miniature full moon." The tail of the comet floated out and away above it like a cone of light, and when the head had disappeared below the horizon, the end of the tail had just reached zenith. Nor was this all. Strange shafts of light seemed to hang straight above the Earth. We seemed to look up through a haze, and it was believed the Earth was actually enfolded in a second tail of the comet. No one, however, felt any discomfort.

It is probable that our atmosphere, thin as it is, would seem like a blanket in comparison with the vaporous particles which make up a comet's tail. The orbits of certain comets lie very close to some of the planets, but they seem to exercise no influence whatever on the planetary bodies. An instance is recorded of a comet coming so near Jupiter that it was actually among his moons. The comet was so upset that he was pulled right out of his old path and set going in a new one. But Jupiter and his satellites showed not the slightest inclination of being aware of the stranger's presence.

Some astronomers think that direful results might occur should we have a head-on collision with a comet. All agree that probably the air and water would be instantly consumed and dissipated, and a considerable region of the Earth's surface raised to incandescence. But yet another consequence equally malign to human interests is foreseen by Professor Todd, who points out that, in the much more probable event of an encounter of the Earth's atmosphere with huge chunks of a large hydrocarbon comet, noxious gases might be diffused in such volume as to render the atmosphere unfit for breathing, and in this way bring death to all forms of animal life.

Comets seem to feel the same fascination for the Sun that moths do for the candle, and those ghosts of the skies which travel in elliptical paths go on wheeling round and round the Ruler of Light like so many gigantic moths. As for those which come in upon the vast parabolas and hyperbolas to pay homage to our Sun, perchance these may make the same obeisance to other suns throughout the stellar space. Certain it is we never see them again, but we know that, like the solar comets, their moth-like fascination for heat and light must at length prove their undoing. For, as a comet approaches the sun time after time "to be invigorated by a good roasting," it must of necessity throw off a good deal of its bulk in the process of tail-making. As it has no possible means of renewing these spent materials, it follows that the comet must certainly reduce itself to a considerable degree at each revolution. The tails, therefore, must decline in size and magnificence. And by and by the tail-making substances having all burned away, we have that sorry spectacle a comet without a tail. We know to a surety that this is exactly what happens. Moreover we have on record proofs of the further disintegration of the comet. Certain comets now travel tandem. It is known that these are the three parts of what was once one huge comet. Other comets have been so broken up that they now make the journey in groups; and in still other cases, final obliteration has so nearly been reached that the comet has literally gone to pieces, existing only in a shower of small stones and fragmentary matter.

In 1832, the world generally had what has since been designated as a "comet scare." Calculators had shown that Biela's comet, which regularly made its pilgrimage to the Sun in a period between six and seven years, was due to sweep over the Earth's orbit. And considerable uneasiness was felt lest the world might be wrecked! In Paris, a regular panic ensued, which was quelled only when the director of the observatory issued a pamphlet explaining the true situation and showing that the comet in its nearest approach to the earth would be at a distance of 50,000,000 miles. Of course, nothing resulted from the appearance of the dreaded ghost! On its next circuit it was not seen, and by the time it was due to appear again almost everybody but the astronomers had forgotten about Biela's comet. They, however, were on the watch, and presently the comet was detected coming on in what seemed a curious pear-shape. All the telescopes were brought to bear upon it, and as the ghostly messenger drew nearer it was seen that the comet had actually divided into two portions. Long before, astronomers had established the theory of the life of comets, but this was the first proof actually accorded to support their belief that the comets began to wane by subdivision.

You may well imagine how eagerly the next appearance of Biela's comet was awaited in astronomical circles! This time the two comets were again witnessed, but the companion comet was now far behind its primary. Indeed, it was coming on so half-heartedly that a million and a quarter miles lay between them! At the next period for return, Biela's comet was unseen, but no one took the matter seriously, for it was so unfavorably placed that detection was very uncertain. The next period, however, was supposed to show the comet up unusually well. Everybody got ready for the visitor, but it failed to appear. The astronomers concluded that it had gone to pieces somewhere out in the vast reaches of space, and no one ever expected to hear anything more concerning Biela's comet.

At its next period, however, in 1872, an extraordinary thing happened. To be sure no one saw anything of the comet, but on the night of November 27, when the earth crossed the trail of the lost one, there was a magnificent shower of shooting stars. Here was the final proof for the men of science! For over five hours the earth ploughed its way through the wreckage of the lost comet. The "rain of fire" was magnificent. Four hundred meteors fell in the small span of a minute and a half; and fire-balls, here and there, apparently as large as the Moon, were observed. At this time, too, a large iron meteorite fell, and was picked up in Mexico. This may or may not have once formed part of the doomed comet.

Sweeping the heavens for comets is an interesting occupation, and the heaviest outlay is in patience and perseverance. Messier, a large discoverer of comets, found all of his with a very ordinary glass, magnifying only five times. Pons, the most successful of all comet-hunters, who has no less than thirty to his credit, began his search while holding the humble position of doorkeeper at the observatory at Marseilles. To-day his name outranks that of the director who kindly taught and encouraged him.

The residue of comets, minute objects, so small that you might carry one in your pocket, if you could pick it up, forms an interesting subject by itself. For these bunches of comet chips, which you may perhaps have regarded as mere rubbish in the great workshop of the skies, have yet another cycle in the marvelous Universe. We shall read about this in the chapter "Shooting Stars."

VII

THE NEBULÆ, OR FIRE MIST

No more remarkable objects are to be seen in the heavens than the hazy celestial clouds known as nebulæ, or fire mist. Likewise, too, they are of the utmost interest, for they give us a clew to the very beginning of the Universe. It is believed that the nebulæ are the star factories from which come suns and the planets and their satellites.

Early astronomers found the nebulæ most puzzling. At first it was the general opinion that these "little clouds," as the term *nebulæ* means, were all star clusters too far away to be seen separately. Herschel, however, advanced the theory that the clouds of nebulous light were not stars, but were made up of huge masses of glowing gas. The invention of the spectroscope proved the truth of this supposition, and showed, moreover, the elements involved. Hydrogen makes up the largest part; helium has recently

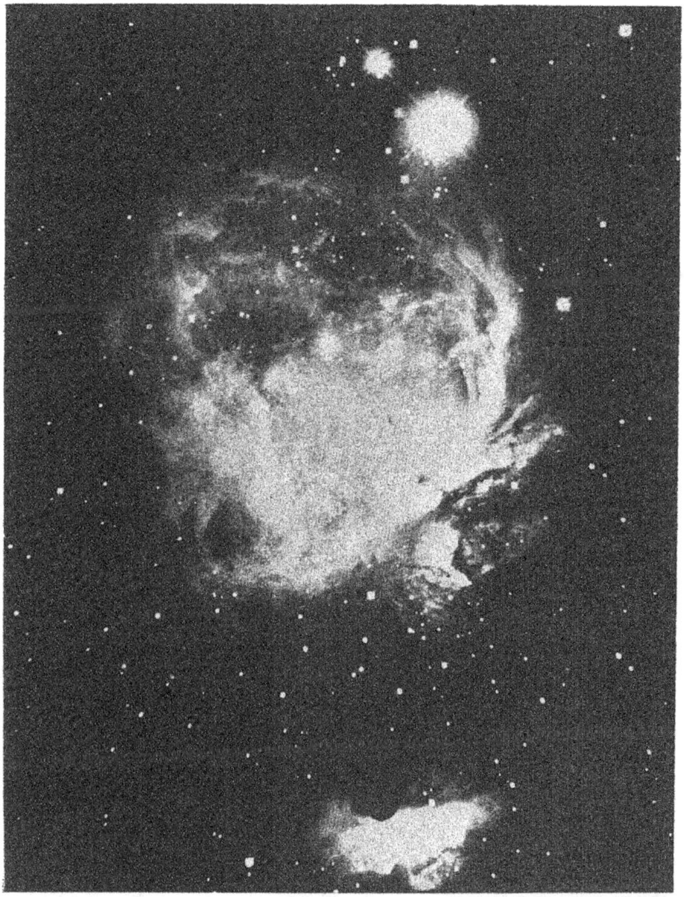

THE GREAT NEBULA IN ORION

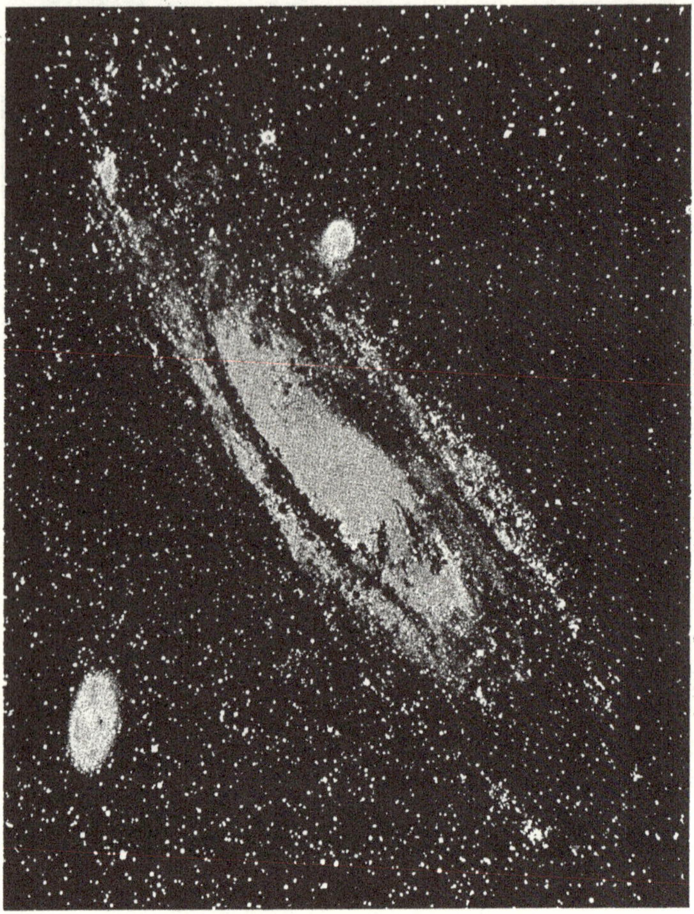

THE GREAT NEBULA IN ANDROMEDA

been added, and there are other substances not yet recognized on the earth.

Two of the most famous nebulæ in the heavens may be observed with the unaided eye. The greatest of these is to be seen in the constellation of Orion, in the winter skies. Some clear night, by looking closely, you will observe that the middle star of the "sword" seems somewhat hazy. Looked at through a small hand telescope, this haziness develops into a considerable cloud against the dark background of the sky. Seen through a large instrument, it becomes a wonderful, wide-spreading nebula, with bright and dark channels, and "extensive wisps of nebulosity" reaching out in many directions, involving other stars. A curious opening, or break in the light, at one side has given it the name of the Fish-mouth Nebula. Another interesting point about the Great Nebula in Orion is that the star which we note with the naked eye resolves under the telescope into not one but twelve stars of various size. Similarly a keen eye will pick out a hazy spot in the constellation of Andromeda, which even the smallest telescopic power resolves into a great nebula. This nebula is much less wide-spread than that in Orion. Though its distance from the solar system is very great, its diameter has been sufficiently calculated to make certain that a great many years would be required even for light to pass from one side of the nebula to the other. It has been estimated that if a map of the Great Nebula in Andromeda could be made, a map of the solar system, drawn to scale, would seem a mere speck if laid upon it.

The known nebulæ in the heavens reach upwards of 500,000. Many of these, of course, are so small, or at such a great distance, that they are only within photographic reach of the great reflecting telescopes such as the marvelous instruments employed at the Mt. Wilson Solar Observatory, near Los Angeles, California. Nebulæ are of all shapes and sizes. To quote a recent English writer, "We have some like brushes, others resembling fans, rings, spindles, keyholes; others like animals—a fish, a crab, an owl, and so on; but these suggestions are imaginative, and have nothing to do with the real problem." In the *System of the Stars* Miss Clerke says: "In regarding these singular structures we seem to see surges and spray-flakes of a nebulous ocean, bewitched into sudden immobility; or a rack of tempest-driven clouds hanging in the sky, momentarily awaiting the transforming violence of a fresh onset. Sometimes continents of pale light are separated by narrow straits of comparative darkness; elsewhere obscure spaces are hemmed in by luminous inlets and channels."

For convenience in classification, astronomers divide the nebulæ into six classes, based on their various forms: (1) annular nebulæ, (2) elliptic nebulæ, (3) spiral nebulæ, (4) planetary nebulæ, (5) nebulous stars, (6) irregular nebulæ, for the most part very large. "If it be realized," says Chambers, "that the word 'annular' is derived from the Latin word *annulus,* a ring, a ready clue will be had as to the general form of the first type of nebulæ." In the constellation of Lyra may be seen, under telescopic power, one of the most marvelous of these gigantic rings of luminous gas. To judge of the size of this ring, Ball tells us that a train, starting from one side of this nebula, and traveling 60 miles an hour, might rush on for a thousand years with unabated speed and then most certainly the journey across this vast immensity would not have ended. "Nor do I venture to say," he continues, "what ages must elapse ere the terminus at the other side of the ring nebula would be reached." The Great Nebula in Andromeda, is a specimen of elliptic nebulæ. "Apparently," says one authority, "it is composed of a number of partially distinct rings, with knots of condensing nebulosity, as if companion stars in the making. Its spectrum shows that it is not gaseous, still no telescope has yet proved competent to resolve it." Spiral nebulæ show such wonderful whorls of nebulosity that they are frequently called "whirlpool nebulæ," and this term seems to describe them fittingly. They are stellar in character, that is like the stars. Their substances are intermingled—gases, liquids and solids. Planetary nebulæ are so called because they show a roundish disc, like the large planets, only much fainter. They are mostly gaseous in composition. Nebulous stars are those hidden in luminous fog. Most of these can be seen only through the telescope.

It is as impossible to measure the distance of

the nebulæ from the Earth, as it is to determine their size. We only know that they are as far away at least as the stars. Nor can we form any definite idea as to what a close-up view would be like. "We can say that the planets are globes like the Earth, with days, nights, seasons, and years," says Macpherson; "We can assert that the stars are suns, like our sun, probably with planets revolving round them; we can even form some idea of what the scene must be at the center of a star cluster; but in the case of a nebula our imagination fails. Their immense size, their enormous distance from our system, and the mighty changes which are believed to be in progress in their midst, show us in a new light the insignificance of the Earth, and increase our astonishment when we remember that only three hundred years ago our little planet was believed to be the center of the Universe."

Another interesting theory is that the nebulæ are the residue of materials of original chaos, from which our solar system and many other such systems as well have been evolved. Certain of the elements found in the Earth are discernible. Those nebulæ which show greenish in color are largely of hydrogen gas. A few of the white ones are resolvable in the high power telescopes into masses of separate stars. For this reason, it was believed in the beginning that all the nebulæ could thus be resolved if only enough telescopic power could be secured.

Strangely enough, the nebulæ seem to be the most numerous in that part of the heavens where there are the fewest stars. It has been found, too, that these vast luminous masses are moving through the heavens at a speed about equal to those of the stars. The bright nebula in Draco, for example, seems coming in towards the Earth at the rate of forty miles per second. On the other hand, the Orion nebula is receding at the rate of eleven miles each second. As yet none of the nebulæ have been discovered to have any axial motion.

There are many difficulties, however, in the way of obtaining accurate details concerning objects so indistinct and remote. Swift as light travels, its rays require at least a thousand years to bring their message from that far-off mass. If at the time of the Norman Conquest of England, the starry clusters of Orion had grown dim, we should but now be finding it out. Through the centuries their diffused light would have persisted like wraiths.

To explain how the stars (the suns of space), and the planets, as well, have been formed, two theories have been advanced, and although these differ in many respects, both begin with the nebulæ. In the first of these, called the *nebular hypothesis,* the idea is that once, far back many hundreds of millions of years ago, all the matter that now makes up the Universe was scattered very thinly through the unlimited vastness of celestial space. Then, as now, the particles of matter attracted one another according to Newton's well established laws. Presently centers of attraction were formed, and these centers drew unto themselves other particles of matter. Thus by the inward falling of matter, and the friction caused by the collision of particles, heat was formed, and the material masses grew into vast nebulæ, which filled all the heavens with luminous fire mist. Millions of years passed, and still the process went on, until finally great nebular whirlpools were set in motion, and began revolving with inconceivable swiftness on their axes. The temperature rose terrifically at centers where condensation became greatest, and presently vast numbers of suns were formed. What followed next we have determined only so far as our own sun is concerned, but there is every reason to suppose that other suns may have developed solar systems by reason of their immense centrifugal force sloughing off great rings of nebulosity which later formed planets, and these in turn their satellites, just as is supposed to have been the case in our own instance, as we have already seen. To be sure, we must bear in mind that all this is merely an hypothesis—that is, a theory. It has never been altogether proved, nor is it likely that it ever will be. However, many great minds have contributed to this theory, and it is the one most universally accepted to account for the scientific development of the Universe. Recent astronomers, however, find this theory still open to serious objections.

The second theory, and more recent one, is known as the *Planetesimal Hypothesis*. This begins with some remote ancestor of our solar system, a more or less condensed central sun, having a slow rotation and surrounded by a vast

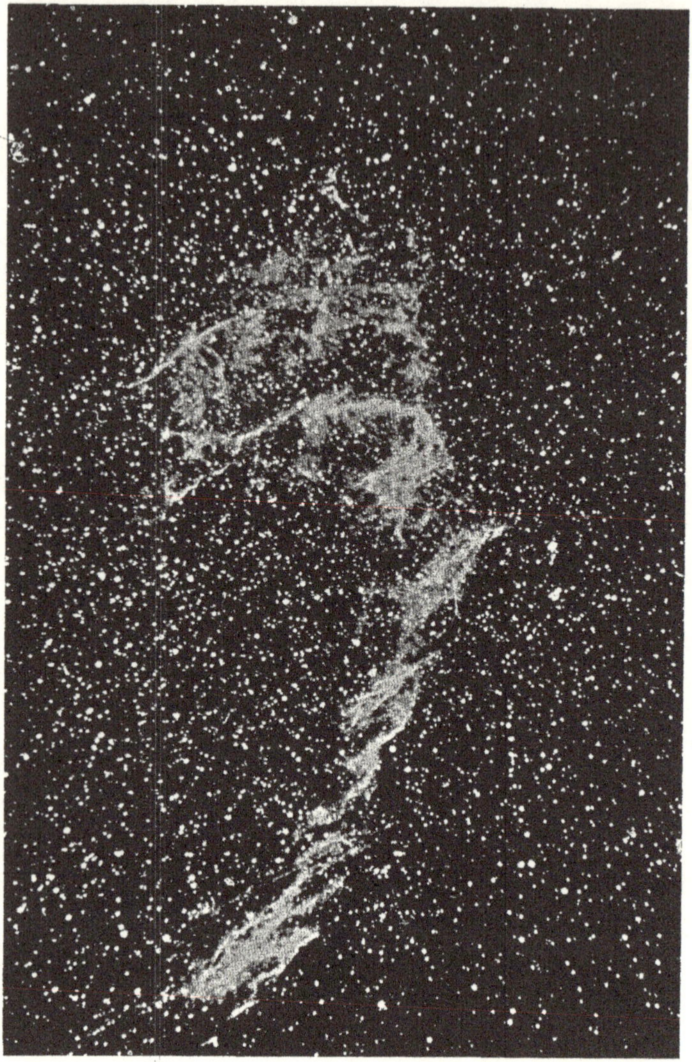

AN IRREGULAR NEBULA IN CYGNUS

swarm of planetoids. All these tiny bodies followed an elliptical path around their sun. The swirling masses seen in the spiral nebulæ to-day are apparently of this type. Now let us see what next happened—if this theory is correct. Out of this great central mass our own sun slowly developed, its body being increased by many of the near-by particles adhering to it, or falling in upon it. Meanwhile, the whirling nebulæ continued to sweep up outlying particles and throw them into globular shape—until they became the planets which in their turn whirled about the Sun. Still smaller particles were swept up and molded into the satellites of the planets, such as our Moon.

The more we reflect on the marvelous wonders and performances of the nebulæ, the more boundless and deathless our Universe becomes. We have gone far in the three hundred years of patiently figuring out our own position in the heavens, but unquestionably we are only on the threshold of the miracles yet to be discerned. There are evidences going to show that our own solar system was once a nebulous mass cool and dark, widely scattered, and revolving but slowly.

Even now the vast spaces of sky may conceal many such masses. Since they do not glow with light we have no means of locating them. We can but speculate. Likewise, too, many of the most familiar nebulæ have grown several times brighter in the years they have been under observation. When we consider that these latter enormous masses of gas, many, many times the area of our own solar system, are sailing on in an endless journey through infinite space, just as the stars and comets are, we can but bow in reverence and awe before the Mighty Intelligence whose hand is on the helm and who alone can answer the queries *Wherefore?* and *Whither?*

VIII

SHOOTING STARS

"Look! Yonder goes a shooting star!" How often we have all heard this exclamation, and followed with momentary interest the little flash of burning sparks which darts along the heavens to disappear in the blackness of space! Perchance, too, we took the incident entirely as a matter of course, giving no heed whatever to the fact that a really, truly star could not "shoot" in this remarkable manner. The stars are suns. They are "fixed" at enormous distances from the Earth, while the phenomenon that we call a shooting star is discharged a few miles above the surface of our planet. Obviously, then, the terms shooting star and falling star are misleading and altogether incorrect; what we really see is a *meteor*, or a *bolide*, as some astronomers term it.

Almost any casual observation of the clear night-time sky will be rewarded by the glimpse of one or more meteors. We should witness the same phenomena during the day, also, were it not for the sunlight. It has been estimated that the dust of 400,000,000 meteors falls to the Earth every twenty-four hours, having a weight collectively of no less than 400 tons. At this rate, it will be seen that the mass of the Earth must gradually be growing larger. Not very noticeably in the course of a year perhaps; but "every little makes a mickle." Our Earth has been hoarding up meteor dust since the beginning of Time. Ages and ages ago it may have been much smaller than it is at present. In fact, as Ball points out, "A large proportion of this globe on which we dwell may have been derived from the little shooting stars which incessantly rain in upon its surface." Moreover, this meteor dust forms no little part of the "dust motes" we see floating in every beam of sunshine. In tropical regions, just before sunset in spring, and before sunrise in autumn, may be seen a pearly radiance arching upward from the sun, in a broad ribbon-like belt. This is called the *"Zodiacal Light."* It is formed from tiny particles of meteoric matter—"diffused dust" held by the attraction of the Sun. Opposite it in the heavens is the counter-glow or *Gegenschein.* Sometimes in European countries this strange meteoric light is seen extending upward like a cone appendage to the Sun.

Usually, if one is observant enough, meteors may be seen falling in groups of twos and threes, now here, now there; and at certain periods of the year, notably in the months of April, August, and November, they occasionally come down in showers, as we shall presently see. Always more meteors fall in the small hours between midnight and six in the morning than at any other time. The reason is very clearly explained by the fact that when one runs rapidly in a rainstorm the chest becomes wetter than the back, for the reason that the advance of the body meets the drops. Likewise, in its revolution about the Sun, the forward part of the Earth is struck by more meteors than any other portion.

The Earth, in this instance, may be likened to a great fisherman. Her air blanket is the net in which the meteors become entangled and thenceforward their career is brief indeed. For countless ages these little bodies—chips from great disintegrating comets—have been whirling through space, continued in the immense orbit of their parent, and covering probably no less than twenty miles per second. We can hardly conceive of such an enormous speed.

If a shooting star should decide to encircle the Earth, it could travel the 25,000 miles and get back to its starting point in a little over twenty minutes. No projectile could be fired from one of our long-range guns that could keep pace with it; and of course the resistance of our atmosphere would be too great for such a projectile.

It is when it reaches the Earth's net that the meteor's "swan song" begins; for this is what its little flash really means. It is the death note of the meteor. Traveling at such an enormous pace, its speed meets with a terrific resistance in the air blanket, heat is kindled, and presently the little particles which make up the meteor turn to gas and flash off in vapor. And then, and then only, do we see the meteor; for it is too small to be detected even by the largest telescopes, and we get no hint of it until at the very moment it begins to be destroyed when its visibility begins. And it is gone in an instant! A wondrous glowing streak—a shooting star in all but reality.

Obviously, on the infrequent occasions when meteors streak the heavens in countless numbers, for hours at a time, as in the meteoric showers of 1799 and 1833, some reason other than the attraction of the Earth must be found. Careful watching showed the astronomers that in the case of every shower the luminous streaks, if prolonged backward, met in a certain small area; that is, they had a radiant point, and, owing to perspective, seemed to fall in parallel lines. It was easy, then, to chart the showers with reference to the nearest neighboring constellation and to give the name of this figure to the meteors. Thus we have Leonids, Perseids, Lyrids, and Andromedids from the constellations of Leo, Perseus, Lyra, and Andromeda respectively. All in all, about 300 radiants are now known, fifty of these being quite well established, but the four mentioned above form the most brilliant showers, and hence are the best known.

Having satisfactorily located the showers, the next thing was to determine their path or orbit and, if possible, to learn their origin. Mathematicians pointed out that the meteors traveled a lengthened orbit, much more like that of a comet than a planet. Then came the miraculous meteoric shower of November, 1833, when from the terrible rain of "fire and brimstone," ignorant peoples everywhere fancied the end of the world was at hand. For nine hours the stars fell "like flakes of snow, . . . varying in size from a moving point or phosphorescent line to globes of the Moon's diameter." Astronomers were quick to note this phenomenon was exactly thirty-four years from the time of a similar shower which had appeared from the same constellation—that of Leo; moreover, it was observed that the apparent orbit of these meteors was the path traveled by Temple's comet. Likewise it was found that the August showers, the Perseids, pursued the same track as Swift's comet (known as the bright comet of "1862 III"), and suspicions grew that comets and meteors had some well-established affinity. The orbit of Swift's comet runs way out beyond the planet Neptune, making a period of 120 years necessary to complete its circuit. Obviously no real proof could be had from it, but in the case of Temple's comet it was not hard to reckon the date for another meteoric display, which Newton fixed as the evening of November 13, and the morning of November 14, 1866, and which subsequently came off exactly as scheduled in a glorious display, well-calculated to delight the hearts of the astronomers. But it was not until the Bielid shower, of November, 1872, which we have already recorded, marking the disintegration of Biela's comet, that the world was satisfied that meteors were no more nor less than the small chips and dust particles of comets, which are thrown off by reason of the burning-out process exercised by the Sun and the attraction of the planets. One other puzzling question yet remained: Why was it that the Earth apparently hit some of these meteors every year, but only once in a certain number of years seemed to run right into the midst of them? This problem, too, was solved, as follows:

"One has only to imagine," says one authority, "a swarm of such meteors at first hastening busily along their orbit, a great cluster altogether, then, by the near neighborhood of some planet, or by some other disturbing causes, being drawn out, leaving stragglers behind, until at last there might be some all round the path, but only thinly scattered, while the busy important cluster that formed the nucleus was still much thicker than any other part. Now, if the orbit that the meteors followed cut the orbit or path of the Earth at one point, then every time the Earth came to what we may call the level crossing she must run into some of the stragglers, and if the chief part of the swarm took thirty-three years to get round (as in the case of the Leonids), then once in about thirty-three years the Earth must strike right into it. This would account for the wonderful display. So long drawn-out is the thickest part of the swarm that it takes a year to pass the points at the level crossing. If the Earth strikes it near the front one year, she may come right round in time to strike into the rear part of the swarm the next year, so that we may get fine displays two years running about every thirty-three years." [1]

It is probable that there are thousands of meteoric currents, comprising the residue of burned-out comets, in our solar system, and

[1] G. E. Mitton, in "The Book of Stars."

large numbers must cross the orbits of other planets, perchance held in check by other suns than ours. Occasionally a chip, manifestly larger than the general run, manages to slip down through our atmosphere without burning itself entirely out, and lodges in the earth, whence it is rescued and termed variously as a *meteorite,* an *aerolite,* or an *uranolith.* Many thousand pounds of such cometary matter have been collected from all parts of the earth, and are carefully preserved in museums, notably in London, Paris and Vienna abroad, and in the United States at the Harvard and Yale Universities, Amherst College, the museums of New York and Chicago, and the National Museum at Washington.

Generally speaking, meteorites are of two classes, meteoritic stones and meteoritic irons, the former being the more numerous. Meteoric stones are encrusted with a thin substance like dense black varnish, caused by the heat generated in their terrific downward flight. Iron meteorites are covered with queer pittings, like deep thumb marks, due to impressions made while white hot by air resistance. The study of meteorites belongs to the province of the chemist and the mineralogist rather than to the astronomer, but so far their analysis has not brought to light any new elements. When seen at night meteorites have a gorgeous appearance. They are like huge fireballs, followed by luminous trains of vapor. In the daytime the light of both fireball and train is largely lost against the deep sky background, and the vapor appears only as a bright cloud. As the solid body hurls along, a deep, continuous roar is heard, ending in one grand explosion, or perhaps in several smaller explosions, and finally the mass may plunge deep into the earth, or it may burst into a number of tiny fragments to be scattered far and wide.

Since the beginning of Time stories of stones and iron missiles hurled from heaven by the wrath of the gods have been current, but it was not until 1803, when a great aerolite fell in France, that scientists began to accept the tales in the light of truth. Then astronomers violently disagreed. Many of them, even such an eminent authority as Sir Robert Ball, held that meteorites were the residue hurled from volcanoes when the world was in the making, and kept in subjection by the Sun, until the Earth came so near them that they could not resist dropping down upon her. Others argued that the projectiles came from the Moon or the Sun. Still others, considering that in only two instances meteorites had been known to fall in meteoric showers, and then with no real proof of probable connection, insisted on connecting them with the planetoids. There were unanswerable objcctions, however, to all of these theories, and at length it was fully determined that the meteorites were in truth large "swarms or shoals of meteoric particles" separated from their parent by vaporization in the rapid journey through space.

Ball estimates that at least 100 grand meteorites fall to the Earth every year. But there are few accounts of eye-witnesses of these marvelous spectacles; nor, on the other hand, do we read of damages caused by the downfall of huge missiles. The largest meteorite on record is probably the immense mass called Ahnighito (the tent), weighing thirty-seven and one-half tons. It was discovered in northern Greenland, by Lieutenant Peary, in 1894. With its two smaller companions, "the Woman," and "the Dog," it had long formed the source of iron for the Eskimos, who held that the three masses had been hurled from the heavens by the evil spirits. "The Tent" was brought to New York in 1897, and now rests under the entrance arch of the Museum of Natural History. Humboldt estimated the diameter of the ordinary large size fireballs from 500 to 2,800 feet. Such balls frequently rival the moon in brightness, and leave behind them a long comet-like train of light. One or two instances have been noted where the train of a fireball remained in full view for half an hour after the meteorite itself had vanished, due it is supposed to phosphorescence.

Amazing as are these great detonating fireballs, darting without warning upon us from the depths of space, as though hurled from the mighty arms of the angered god of the thunder, they yet find their rivals in sheer magic in their small kindred, which weigh at best only a few ounces. These latter are truly the final marvels of the solar system. They show us our Sun in a new light. We see him as a condescending ruler, mindful of even the tiniest object in his kingdom—the shooting stars. Whether of the size of a bullet, a common rifle shot, or even a tiny grain of sand, he holds them in an elliptic course around him as carefully as he does the great king of the planets, Jupiter himself. Whole shoals of infinitesimal meteors are inspired by his gentle might to pursue their miraculous journey in one common purpose, each one pushing on serenely in its path, independent of its neighbors, round and round, year in year out, content with its lot of allegiance, until, presto! it is whisked from the path by the force of Mother Earth's alluring charms, and perishes in a final picturesque flash of light. "The soul of a departed one," the Indians murmur softly, when they see it.

IX

COLORED AND DOUBLE STARS

No doubt if any one asked you the color of the stars, you would answer *white* without an instant's hesitation. But, suppose you carefully scan the heavens the first clear night: you can not help seeing that some stars show reddish or yellowish tints, others shine with a steely blue light. Looked at through a telescope these stars resolve themselves into brilliant individual colors: the red stars are bright red, copper red, blood-red, "glowing like a live coal," etc. Likewise the blue, green, and yellow stars shine out in various tints and shades of the most vivid types. There is a message in this wealth of color, which the spectroscope readily aids the astronomer to read. It tells him of what elements these various suns are composed: thus the red stars are largely carbon; the blue are hydrogen; the yellow ones show a mixture of elements like our Sun; and so on.

But the most interesting thing about these colored stars is that they are nearly always double; that is, they are made up of two stars so close together that the naked eye is unable to separate them. At first astronomers could not believe what the glass told them. They thought that these stars seemed to be doubles only because they happened to be nearly in the same line of sight from the earth. It was supposed that one star might easily be many millions of miles in space behind the other. Several such pairs of stars were already known to exist in the heavens; astronomers termed them *optical doubles.* Certain reasons, however, argued against classing the colored doubles with these stars, and the problem stood unsettled until 1802, when Sir William Herschel proved that the puzzling doubles were in truth just what they seemed. They were saved from coming together by mutual attraction and ending in an inglorious crash—the chief argument against believing in two stars so closely associated—because both were in motion, one revolving around the other. They were confined at certain distances by the laws of gravitation, just as our Sun and the planets are. To distinguish this class of doubles from those which merely seemed to be doubles, Herschel suggested calling them *binary stars,* and this name still stands.

Many of the brightest stars in the sky are binaries. Alpha-Centauri, our nearest fixed star, you remember, is included among these. It is made up of two very bright stars, which take eighty-one years to travel round their orbit. At their closest point they come as near together as Saturn is to the Sun—886 millions of miles; when farthest their distance is far beyond that of Neptune from us. Sirius, the brightest star in the sky, and Castor, one of the well-known "Twins" in the constellation of Gemini, are two other doubles easily observed by the smallest telescope.

Castor is one of the most familiar of the double stars, and astronomers long ago discovered that one star is revolving slowly about the other—so slowly that several hundred years are required for a complete circuit. This is not so surprising when we remember that Castor is so remote from us, that the two stars instead of almost touching each other are really hundreds of millions of miles apart. If, for illustration, we could be upon one of these gigantic worlds observing our own solar system in the heavens—and if one of our outermost planets, say Neptune, should be enlarged to a size near that of the Sun—then Neptune and the Sun would form a double star, slowly revolving the one about the other. The intermediary planets would be invisible, unless some extremely powerful Castorian glass should pick up Jupiter, as a tiny satellite hovering near.

Upwards of 12,000 double stars have been counted in the heavens, and the orbits of many of them successfully calculated. Strangely enough, many of these doubles were discovered simply by inference, irregularities in certain stars having suggested the influence of another body, just as the action of Uranus led astronomers to suspect the existence of Neptune long before they were actually able to prove his existence. In some instances triple, quadruple, and even more stars have been discovered banded together in this miraculous fellowship. "These multiple systems vary from one another in almost every case," says Mitton. "Some are made up of a mighty star and a comparatively small one;

others are composed of stars equal in light-giving power—twin suns. Some progress swiftly round their orbits, some go slowly; indeed, so slowly that during the century they have been under observation only the very faintest sign of movement has been detected; and in other systems, which we are bound to suppose double, the stars are so slow in their movements that no progress seems to have been made at all."

An especially interesting feature about double stars is that the two partners are often of contrasting colors. The most beautiful example within range of the ordinary telescope is Beta-Cygni in the constellation known as "The Swan." The larger star is reddish-yellow and the smaller one sapphire-blue. Antares, already mentioned as a fiery red star in Scorpio, one of the Zodiac constellations, has a small green companion. Other double stars show pairs of "yellow and rose-red, golden and azure, orange and purple, orange and lilac, copper-color and blue, apple-green and cherry-red, and so on. In the Southern Hemisphere there is a cluster containing so many stars of brilliant color that Sir John Herschel named it "The Jeweled Cluster." "Here are to be seen companion suns in cream-white, rose-color, lilac, russet, fawn, buff, and olive hues in endless numbers.

Imagine our sun sharing his kingdom with a royal purple companion! But stay, could our planets exist under the pull of two suns in opposite directions? Astonishing as these suggestions seem, it is certain that conditions even more amazing exist in some of these double stellar systems. For, besides these glowing suns, they contain huge dark bodies which may very well be planets. Indeed, says one authority, "In some cases the dark body which we cannot see may even be larger than the shining one, through which alone we can know anything of it. Here we have a new idea, a hint that in some of these systems there may be a mighty earth with a smaller sun going round it, as men imagined our sun went around the earth before the real truth was found out." In the famous quadruple system of Zeta-Cancri three bright stars are supposed to revolve round a dark body, which appears to be by far the largest of the four.

So far as we know, many of these stellar bodies may be the homes of human beings. If so, what an endless variety of celestial sights must delight the eyes of a dweller in this part of the Universe! Proctor, in his interesting work, *The Expanse of Heaven,* pictures a world where twin suns, one blue and the other orange, rise together to produce "double day," or perchance as the orange sun sets the blue one rises, and there is no night. "The skies must be exceedingly beautiful," he tells us. "Our clouds have their silver lining because it is the light of the Sun which illumines them. Our summer sky presents glowing white clouds to our view, and at other times we see the various shades between whiteness and an almost black hue. . . . But imagine how beautiful the scene must be when those parts of the cloud which would otherwise appear as simply darker shine with a fuller blue light or with a fuller orange light. How gorgeous again must be the coloring of the clouds which fleck the sky when one or other sun is setting."

From double and multiple stars it is but a step to groups and clusters of suns, and here again we have a picture of another system of worlds. The Pleiades or Seven Sisters is the most famous star group or cluster. It is a noticeable object in the winter skies, and under the telescope resolves into some five or six hundred stars, which are in turn multiplied fivefold on the resourceful photographic plate. Another interesting cluster is that of "the Beehive" in the constellation of Cancer. But perhaps the very finest star cluster in the whole heavens is that known as "Messier 13." This is the famous cluster in the constellation of Hercules. It cannot be seen as a cluster to the naked eye. It is just visible as a faint star. Yet studied through powerful lenses it is found to contain at least 50,000 stars which are estimated as 200 times brighter than our Sun. "The total number of stars in this cluster, as in others of the globular type," says Serviss, "cannot be counted, because of their increasing density towards the center, but reasonable estimates show that there must be hundreds of thousands and possibly a million of them, the most of which are giants compared with the Sun." Seen through a telescope for the first time, the amateur, according to an old Scottish astronomer, "cannot refrain from a shout of wonder."

Imagine what it would be like to live on a planet situated in the middle of a star group such as the Herculean cluster! Such a world would be bathed in perpetual day. One sun after another would blaze its fiery way across the heavens; and if by chance the larger ones should give way to a semblance of night, then their sky would be brilliantly pointed by countless stars of the first magnitude. But it is doubtful if their sky would ever be dim enough to see even these. Their astronomers would be in utter ignorance of the stars and planets as we know them.

In thinking of the stars as suns, naturally we have pictured them as suns of the same model as our own, but we see now how erroneous such a supposition would be. Indeed, there are so many different groups and double suns in the heavens, that it is possible that solitary suns such as ours are the exception and not the rule in the vast reaches of the stellar universe. More and more it is borne in upon us that the world in which we live is but one kind amid an infinite variety of worlds. We are lost in the contemplation of a Universe without bounds, a Universe in which our Earth sinks into "an absolutely insignificant atom."

X

THE MILKY WAY

Pure leagues of stars from garish light withdrawn
  Behind celestial lace-work pale as foam,—
I think between the midnight and the dawn
  Souls pass through you to their mysterious home.
—*William Hamilton Hayne.*

MILTON also spoke of the Milky Way as: "The way to God's eternal house." The Norseman saw it as the path to Valhalla, over which traveled the souls of heroes who fell in battle. In *Hiawatha,* we are told how Nokomis taught the little Indian lad about the stars that shine in the heavens:

Showed the broad white road in heaven,
Pathway of the ghosts, the shadows,
Running straight across the heavens,
Crowded with the ghosts, the shadows,
To the Kingdom of Ponemah,
To the land of the hereafter.

In Sweden the peasantry speak of the Milky Way as the "Winter Street," and Edith M.

Thomas has woven this thought into some beautiful verses, beginning:

> Silent with star-dust, yonder it lies—
> The Winter Street, so fair and so white;
> Winding along through the boundless skies,
> Down heavenly vale, up heavenly height.

Ancient peoples of various races have likened the Milky Way to a broad river. It was into this stream, the river of heaven, that the burning chariot of the Sun was plunged on the occasion of Phaëton's mad drive. Our English ancestors often spoke of the Milky Way as "Jacob's Ladder." So we might go on recounting one symbol after another that has been suggested to different peoples since time began. But grander than any of these fanciful thoughts is the real truth of this magnificent arch across the zenith of our night-time skies. It is a mighty circle of light, science tells us, "composed of worlds heaped on worlds, suns towering beyond suns, in a profusion that startles the imagination and awes the soul."

For centuries the *galaxy*, as this mysterious circle is scientifically termed, has been an object of close interest among the astronomers.

THE MILKY WAY AROUND THE STAR CLUSTER, MESSIER II.

Aristotle thought that it might be due to atmospheric vapors. Another early student inclined to the absurd notion that the galaxy was the shadow cast by the Earth on the heavens. Soon, however, scientists began to feel sure that it was a broad path of stars too far away to be separately seen, and when Galileo turned his newly-invented telescope upon it, this supposition was found to be the true one. But Galileo's telescope was far from powerful enough to resolve the galaxy. Certain individual stars stood out quite clearly, but back of these was the same puzzling field of misty light, suggesting a breadth and depth of the stellar universe hitherto unguessed. Subsequent telescopes and photographs have still left many things about the galaxy unsolved. As Mr. Gore most fittingly observes: "The Copernicus of the sidereal system has not yet arrived, and it may be many years or even centuries before this great problem is satisfactorily solved."

We do know, however, that the Milky Way is a mighty star stream—that is a system of stars which seem to us to be connected, but which may yet be separated by millions of miles. All are mighty orbs, suns of the same type as our own Great Ruler. Many of them vastly larger than he, others smaller. Each of them, we are certain, possible centers of planetary systems, and the homes of human beings. Small indeed would be our own place in the Universe if viewed from this celestial highway. In truth, neither our Earth nor any of the other planets in our solar system could be seen. Our Sun, if visible at all, would be nothing but a faint, a *very* faint star. Indeed, its light might be extinguished, and all the human race, as we know it, wiped out, without causing any stir in that distant part of the Universe. Professor Ball tells us that in this event: "All the stars of heaven would continue to shine as before. Not a point in one of the constellations would be altered, not a variation in the brightness, not a change in the hue of any star would be noticed. The thousands of nebulæ and clusters would be absolutely unaltered; in fact, the total extinction of the Sun would be hardly remarked in the newspapers published in the Pleiades or in Orion. There might possibly be a little line somewhere in an odd corner to the effect that 'Mr. So-and-So, our well-known astronomer, has noticed that a tiny star, inconspicuous to the eye, and absolutely of no importance whatever, has now become invisible.' "

The galaxy extends roughly in a great circle entirely around the celestial sphere. Usually about half of it is to be seen above the horizon on clear moonless nights, like a great belt of uneven haziness, stretching from zone to zone, and as broad across as three full moons. Between Centaurus and Cygnus, it divides into two branches. Miss Clerke, in her *System of the Stars,* says: "Involuntarily the image presents itself of a great river, forced by an encounter with a powerful obstacle to throw its waters into a double channel, lower down merged again into one. The intervening long strip of islanded rock and gravel might stand for the great rift between the branches of the sidereal stratum, which, although to the eye, owing to the effect of contrast, darker than the general sky, is in reality nowhere quite free from nebulous glimmerings. It is encroached upon by fringes, effusions, and filaments, spanned by bridges of light, and here and there half filled up by long, narrow, disconnected masses or pools of nebulæ, lying parallel to the general flow of the stream."

Photographs and telescopic views of the Milky Way show remarkable rifts and chasms which until just recently were regarded as glimpses of the depths of space beyond. Now scientists have concluded that this "darkness behind the stars" is, in truth, not space at all, but that it marks the presence of non-luminous bodies. According to Professor Barnard, of the Yerkes Observatory, they are simply "cloud masses of dark nebulæ whose light has failed them, or which never had any light." One of these apparent rifts in the Milky Way may be plainly noted with the naked eye. It has been known to navigators for ages as "the coal sack," and is about eight degrees long by five degrees broad. Only one star may be marked in this space unaided, but several show up under the telescope.

The Milky Way has been termed "the ground plan of the Universe." It has long been noticed that there are more stars in the region of the heavens near to its broad path than there are in the opposite direction. In short, the stars increase up to the galaxy, which seems to be a broad highway for stellar clustering. It is apparently the equatorial zone of the stellar universe. The stars here appear to be closer together than in any other part of the heavens, but this may be due to circumstances affecting our line of vision. Since this celestial path arches about us, it is not unlikely that our Sun is himself a member of the Milky Way. Sir John Herschel, who made a study of this portion of the stellar system from both hemispheres, inclined to the belief that "our situation as spectators is separated on all sides by a considerable interval from the dense body of stars composing the Galaxy." Time, perhaps, may satisfactorily solve this stupendous problem, but at present nothing seems more unlikely.

A recent investigator likens the general shape of the stellar universe to that of the Great Nebula in Andromeda. Furthermore, scientists have argued that, while space is undoubtedly boundless and infinite, there is a definite number of stars or suns, possibly 500,000,000 or more. Were the number of stars as infinite as space itself the whole heavens would shine with a brightness equal to that of our Sun. But, even granting that the stellar system may have boundaries in space, it is yet of a vastness entirely outside the comprehension of man. Supposing the Milky Way to be the middle zone of the stellar universe, mathematicians have calculated that the nearest sun at the outer edge of this broad path is at a distance that must be expressed way up in the seventh family of numericals, thus 000,000,000,000,000,000,000. Any attempt to realize the immense distance marked by such a speculative guide post sets our heads swimming! Much less, then, can we essay to measure the further breadth of this marvelous stellar space. And yet this is but a beginning. If we would comprehend the *whole* Universe, astronomers say that there may be other stellar universes out in illimitable space; that is, what we know as the star world may have neighboring star worlds, and these their neighbors away out in the Infinite Beyond.

"The number of stars and systems really existing, but invisible to us, may be practically infinite," says Mr. Gore. "Could we speed our flight through space on angel wings beyond the confines of our limited universe to a distance so great that the interval which separates us from the remotest fixed star might be considered as merely a step on our celestial journey, what further creations might not then be revealed to our wandering vision? Systems of a higher order might then be unfolded to our view, compared with which the whole of our visible heavens might appear like a grain of sand on the ocean shore—systems perhaps stretching to Infinity before us and reaching at last the glorious 'mansions' of the Almighty, the Throne of the Eternal."

XI

THE CONSTELLATIONS

Mark Twain once observed that the thing that puzzled him the most about astronomy was how we found out the names of the stars. Perhaps you, reader, may have wondered about this very point yourself! Evidently imagination played a large part in christening the stars, for the heavenly groups, or constellations, as we call them, bear the names of figures and animals, for the most part so wildly fanciful that it is impossible to conceive how any one could have gone so far out of the way to imagine anything so ridiculous. And yet, beneath these figures, wild and fanciful as they are, lie many beautiful and ennobling ideas, which are of interest not only for their real merit, but as sidelights on the earnest efforts made by the early peoples to understand what was to them the most appealing problem—the relation of the Earth to the heavens.

The word *constellation* is made up of two Latin words: *con,* which means *together,* and *stella,* meaning *stars.* In plain English, then, the word *constellation* means *stars together.* Almost all nations have arranged the stars into constellations, but it is supposed that the Greeks mapped out the geography of the heavens which is now used, 1,400 years before Christ, drawing largely on the records of the gods and heroes associated with the voyage of the fabled ship *Argo.* That they also followed the leading of the Chaldeans and the early Egyptians is proven by the fact that no less than forty-eight of the constellations, numbering all of the largest and best known groups, had already been recorded by Ptolemy, more than 600 years previous. Therefore, the actual date of the invention of a name for even one of these popular groups is not known; we only know that they have been thus called so long that, as the legal lights would say, "the memory of man runneth not to the contrary." Consequently, though we find it difficult to picture the likeness of a bear, a bull, a fish, the most agile of hunters, or what not, in the scattered groups of stars which make up the eighty or ninety constellations now catalogued in the stellar world, we still cling to the old names as a matter of convenience in referring to them.

If the axis of the earth were straight up and down in regard to the plane of the Earth's orbit round the Sun, then we should always see the same set of stars, regardless of time and season. but as the axis is tilted slightly, in winter in the Northern Hemisphere we see more of the sky to the southward than we can in the summer; while in the Southern Hemisphere far more stars to the north can be seen at this period. Always, however, there is one fixed point in each hemisphere round which the other stars seem to swing. This is the point directly over the poles. Happily, too, there is a bright star just where the North Pole would seem to touch the sky could it be drawn outward. This is the Pole Star. For ages it was the mariner's only compass:

> "Coasting, they kept the land within their ken,
> And knew no north but when the Pole-star shone."

Bryant in his *Hymn to the North Star* thus apostrophizes it:

> The sad and solemn night
> Hath yet her multitude of cheerful fires;
> The glorious host of light
> Walk the dark hemisphere till she retires;
> All through her silent watches, gliding slow,
> Her constellations come, and climb the heavens, and go.
> And thou dost see them rise,
> Star of the Pole! And thou dost see them set.
> Alone in thy cold skies,
> Thou keep'st thy old unmoving station yet,
> Nor join'st the dances of that glittering train,
> Nor dipp'st thy virgin orb in the blue western main.

Do you know how to mark the position of the Pole Star in the sky? First, you must learn to recognize the Big Dipper. Perhaps this has already been pointed out to you. If not, you will have no difficulty in locating the seven bright stars which form this conspicuous object. Of course, the position of this constellation, like that of every other in the heavens, varies with the hour of the night and the season of the year, but in the Northern Hemisphere it is always to be seen somewhere in a clear sky. In April, at eleven o'clock at night, it is high overhead, and appears upside down. In September, at the same hour, it is low down in the north, seeming to rest upon the horizon in an upright position. In July, it is to be found in the west, and at Christmas time, in the early evening hours—the most favorable time for observation—it is in the east. The two stars on the side away from the handle of the Dipper are called the Pointers. A line drawn about three times as far as the distance between the two, in the direction they indicate, will lead directly to the Pole Star.

The Big Dipper is the constellation which the ancients designated as the Great Bear. In England, it is sometimes called Charles's Wain, or Wagon; another name for it is The Plow. The Pole Star forms the last star in the tail of the Little Bear or Little Dipper, this constellation is very like the Great Bear, but so much smaller and fainter that it is considerably harder to locate. The Little Bear is turned the opposite way from the Great Bear and its tail points in the opposite direction. A large amount of imagination is necessary to see a bear in either of the figures formed by these constellations. And who ever heard of a bear with a tail! A fox or a dog or almost any other

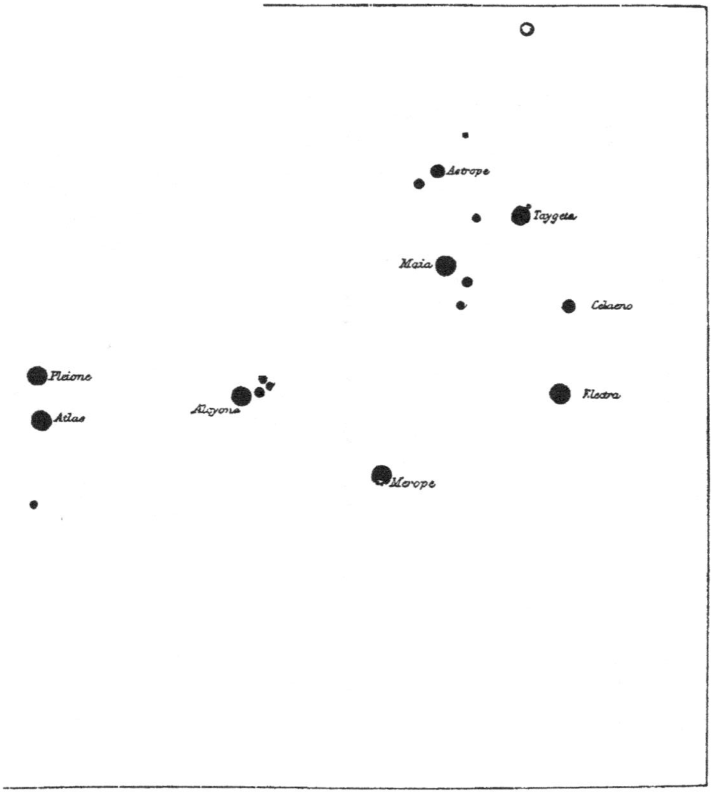
Astrope
Taygeta
Maia
Celaeno
Pleione
Alcyone
Electra
Atlas
Merope

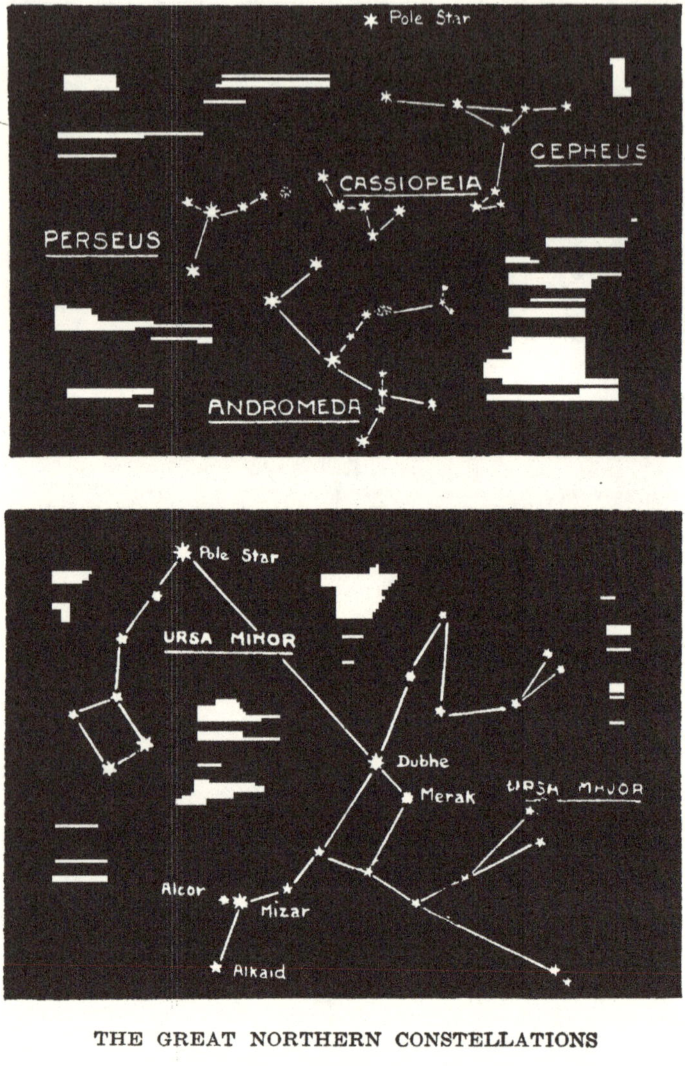

THE GREAT NORTHERN CONSTELLATIONS

animal would fit the vague outlines better. Ancient peoples, however, of nearly all races, saw a bear. Homer spoke of it keeping watch upon Orion from its Arctic den, and there are countless references to it in both classic and modern literature.

> One after one the stars have risen and set,
> Sparkling upon the boar-frost of my chain;
> The Bear that prowled all night about the fold
> Of the North-star hath shrunk into his den,
> Scared by the blithesome footsteps of the Dawn.
>
> —Lowell—*Prometheus.*

Mythology recognized in the constellation of the Great Bear the beautiful Callisto, who unfortunately excited the jealousy of Juno, and was changed into a bear by the angry goddess. While wandering in the forest the bear, it was said, met with her own son and sprang to embrace him. The hunter, all unknowing, raised his spear to strike, and Jupiter in pity snatched both into the sky, where they became the Great Bear and the Little Bear. Juno, however, was still very wroth, and not to be cheated entirely of vengeance she warned Oceanus never to let them come near his watery domains. Hence the

Bears must ever wander round and round the pole, nor venture to dip their huge bodies beneath the horizon. This condition, perforce, gives way in the equatorial regions, for there constellations which are circumpolar in our latitude begin to rise and set. Thus the poet accounts for this circumstance:

> We saw the Bears, despite of Juno, lave
> Their tardy bodies in the boreal wave.

Astronomers speak of the Great Bear and the Little Bear as Ursa Major and Ursa Minor, which while very high-sounding and scholastic is no more nor less than the two simple terms rendered in Latin. Zeta, the middle star in the tail of the Great Bear, is a famous double, which resolves under a small telescope into two wonderful stars, commonly known as Mizar and Alcor. A sharp eye can detect these doubles unaided. In European countries, these stars are referred to as "the horse and his rider" or "Jack on the middle horse," following the idea that this constellation is a plow drawn by three horses. In German mythology Alcor becomes "Hans the Wagoner," who in return for aiding the weary Savior was offered the kingdom of heaven, knowing himself unworthy, however, he begged permission instead to drive the celestial plow horses, and since time out of mind has been seen astride the middle horse. Job spoke of "the Bear with her train," thus sensibly seeing in the three stars a following of cubs, rather than a tail on an animal that has always been tailless. Our own Indian tribes pictured the three stars as a hunter and his dogs following the trail of the stellar bear in a wondrous chase which lasted from early spring till autumn, when the animal was wounded and its blood sprinkled the leaves to crimson, russet and brown.

The Big Dipper and the Pole Star constitute the "Great Star Clock of the North," which has guided fleet and caravan over wastes of sea and sand since Time immemorial. The Pole Star is the center of the great clock face, and the Pointer stars form the hour hand. Due north on the horizon is 12 P. M. Each quarter-circle is six hours, and because the Big Dipper seems to swing round from *left* to *right* the pointers run counter clock-wise. With a little practice at drawing imaginary clock-faces round the Pole, you will be able to gauge the time within a half hour, possibly less.

A line extending about as far beyond the pointers as they are from the Pole Star will show, on the opposite side of the Great Bear, a large capital W, made up of five or six stars, shining so brightly that you will wonder how it is that you have never noticed this constellation before. This is Cassiopeia. If you look sharply, you may perhaps see the lady in the chair which the ancients fancied they saw in this group. She is the Ethiopian queen who dared to set her beauty above the sea-nymphs, and was sentenced by the gods to be bound in her chair and swung into the heavens, where she might revolve around the Pole, now head upward now downward, in order to teach her humility. In looking at Cassiopeia, you will note, of course, that this constellation seems to be on or near the great belt of the Milky Way, with which we are already familiar.

The Pole Star, the Dippers, and Cassiopeia constitute what is called the *Great Northern Constellation,* and serve as convenient celestial landmarks, which are always at hand, for mapping out the heavens. You will find it interesting to trace out by their help others of the best known groups in a personally conducted tour of your own. A good star atlas would make things a bit easier perhaps, but you really do not need it, if you are good at executing imaginary lines and angles. Above all, avoid trying to locate the constellations by the aid of maps outlined in the grotesque figures fancied by the primitive people. These only lead to confusion. Such figures never existed save in the richest imaginations, and one should never attempt to picture them in the sky until after the group with which they are associated is well-known. They are interesting only as side-lights on the beginnings of a subject which is apt at times to become too solid, and they have absolutely no place in any real study of the geography of the sky.

Before going on with the constellations, we must prepare ourselves to understand the designations used by astronomers to indicate the magnitude or brightness of the stars. In general, small letters of the Greek alphabet are used to denote the most prominent stars of a group. α represents its brightest star, β the next, γ the third, and so on. As a further identification, the Greek letter followed by the Latin genitive of the constellations shows to what group it belongs: thus α Orionis is the brightest star in Orion, γ Andromedæ is the third star in order of brightness in Andromeda. If a constellation has more than twenty-four stars deserving especial mention, then the letters of the Latin alphabet are used: if these prove insufficient, ordinary Arabic numerals follow. Thus we have *f* Tauri, 61 Cygni, and the like. Further than this, about one hundred stars have proper names, mostly of Arabic origin. These stars usually may be expressed in two ways; for example, α Tauri is none other than Aldebaran, the well-known Bull's eye, α Lyræ is the bright star known as Vega; β Ursa Majoris is Merak in the constellation of the Great Bear (the bottom one of the Pointer stars).

Because of the rotation of the earth on its axis, all of the constellations except those of the Great Northern seem to cross the sky from east to west, rising and setting four minutes earlier each night. This slow and constant ever-changing procession thus brings to each season its own constellations, so that star-gazing never loses in interest. Always the scene is ever new and yet ever old, filled with the fascination of what are to us new discoveries and with the welcoming of old friends: in spring, Gemini (The Twins), Leo and Virgo; in summer Hercules, Scorpio and Boötes; in Autumn, Cygnus, Pisces, and Aquila; in winter, Orion, Perseus, Canis Major (Sirius) and Taurus (the Bull).

One of the most conspicuous figures during autumn and winter is that of the Great Square of Pegasus—the final home of the famous winged horse of ancient Greece, whose marvelous doings are doubtless familiar to all, since they are so charmingly told in Hawthorne's *Wonder Book*. To reach this Square we draw an imaginary line from the Pole Star over the end of Cassiopeia and as far again, when we come to four stars which if enclosed, would suggest to our modern minds a big saucepan, instead of the famous horse, and on looking about we find that this symbol is quite complete, for there is a bent handle attached to the pan! However, this is the Great Square of Pegasus, winged horse or not, and it is located in one of the most interesting tracts of Star-land. Half-way between one corner of the square or "pan" and Cassiopeia is the Great Nebula in Andromeda. The star in the handle is an exquisite orange-colored double with a sea-green companion. Near the end of the handle is Perseus. Some of you may perhaps recall the tale of this hero's marvelous quest in seach of the Medusa's head. When he and Andromeda, the daughter of Cassiopeia, whom he rescued from the cruel fate to which her mother's sins had bound her, were snatched up into the sky, there to regain peace, Perseus triumphantly bore with him the Medusa's head. Algol, the wonderful variable star, is the "baleful, blinking demon-eye" of this terrible head, "which about every third day drops from the second magnitude to the fourth and recovers in a few hours." The reason for this astonishing diminution of brightness is attributed to the fact that Algol has a dark companion, about the size of our Sun, which in the circuit of its orbit manages to come in between the star and the observer and partially cuts off the light; Algol himself being a star about one million miles in diameter.

A sharp curve to the left from Perseus draws the eye to Capella, "the goat" in the wide-spreading constellation of Auriga, the Charioteer. Capella is a bright star of the first magnitude, which we already know as a sun four thousand times larger than our own brilliant orb. Not far from Capella is an odd arrangement of three stars in a triangle, the "Hœdi" or three kids, so the fertile minds of the ancients termed them. Capella was supposed to nurture these frisky animals, who were held in such evil repute that a poet in the third century B. C. counseled the mariners,—

> Tempt not the winds, forewarned of dangers nigh,
> When the Kids glitter in the western sky.

In late autumn and winter Capella rises high up in the sky, and then there may be seen below her and somewhat to the westward the famous Pleiades, or Seven Sisters, only no one has ever seen more than six stars with the naked eye. In the old days people attached particular good luck to the number seven; so it is quite possible that those who named this constellation invented the tale of the lost Pleiad, which has cropped up in literature since time out of mind. An old writer tells us that the Berbers and Dyaks located the center of the Universe and the abode of the Deity in the Pleiades. "With November, the Pleiad month"—when this constellation reaches its highest pinnacle in the southeastern heavens—"many primitive people began their year; and on the day of the midnight culmination of the Pleiades, November 17, no petition was presented in vain to the ancient kings of Persia. The same event gave the signal at Busiris for the commencement of the feast of Isis. . . . Savage Australian tribes to this day dance in honor of the 'Seven Stars,' because 'they are very good to black fellows.'" Alcyone, the brightest star of the group, reaches its best period about the time of the winter solstice, when all the world is sunk for a few days in restful calm. Sometimes this is a "golden cluck-hen" with her brood of chickens about her;

AURIGA
Capella
The Goat
The Kids

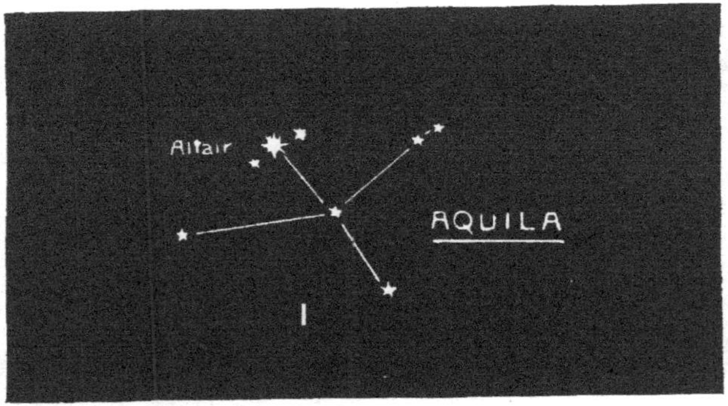
Altair
AQUILA

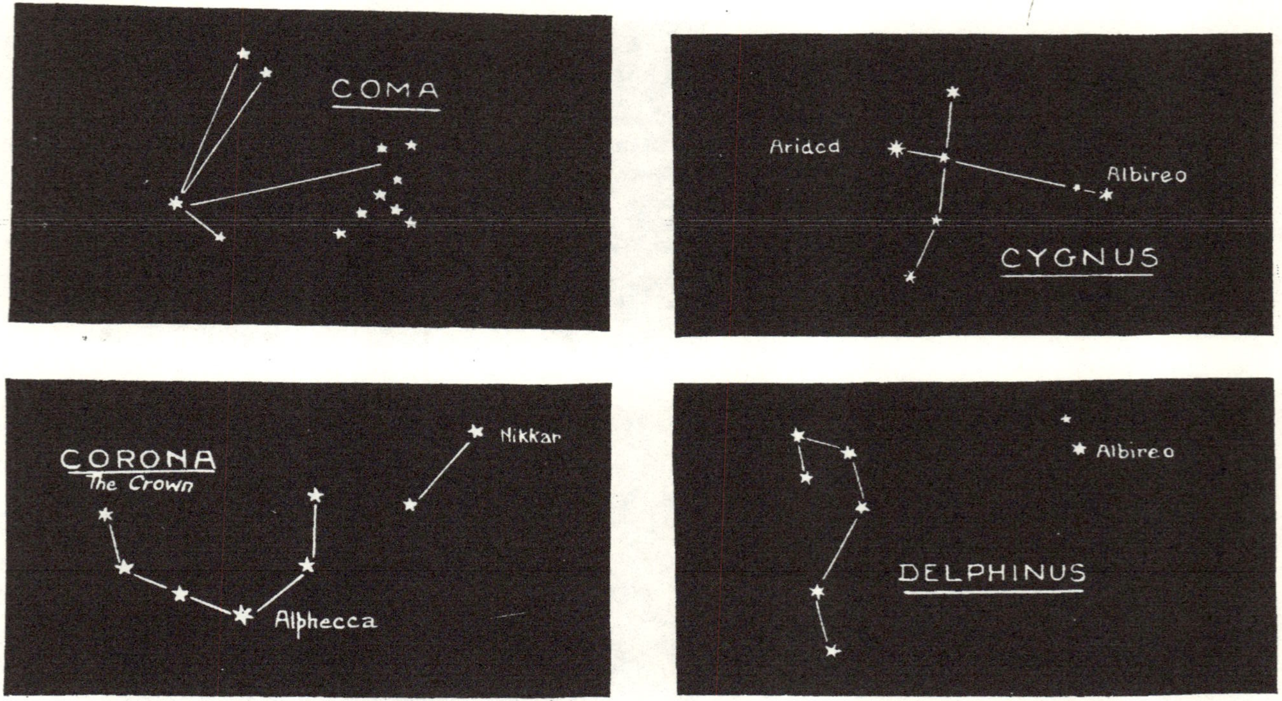

CONSTELLATIONS II

again it is a girl feeding her flock. Classical poets have fancied the Pleiades as a flock of pigeons fleeing from Orion, the hunter. Bayard Taylor styles them as "golden bees upon the mane of the bull." While Tennyson likens them to a glittering "swarm of fire-flies tangled in a silver braid." To the Greeks, they were sailing stars, their rising in May betokening the time for the opening of navigation, hence the name *Pleiades,* derived from their verb, meaning "to sail." Owing to what astronomers term the *precession of the equinoxes*—which is literally a gliding backward or westward of the equinoxes, about fifty and one-fourth degrees annually, due to the slow movement of the equator round the ecliptic,—the Pleiades no longer rise in May. They are not seen until about September 1, when they appear low down to the east. By the first of March this interesting group has apparently progressed around to the western sky and may be seen setting about 11 P. M.

As the Pleiades begin to mount into the heavens, a glorious reddish star of the first magnitude rises beneath them, and a little to the southeast. This is Aldebaran, which is already well-known to us. It is the bright star of the Hyades group, which mythology terms half-sisters of the Pleiades, and with them forms a part of the constellation of Taurus the Bull. The Hyades are described by the poet as "whitening all the Bull's broad forehead." The name comes from the Greek word rain, and when these stars are close to the horizon "a spell of weather" may be expected. "In the showery springtime," says Porter, "they set just after the sun, and in the stormy period of late fall just before sunrise. . ." The classic writers again and again refer to them as the rain stars; Spenser called them "moist daughters," and in Tennyson's *Ulysses* we read:

> Through scudding drifts the rainy Hyades
> Vex'd the dim sea.

The glory of our winter skies is the constellation of Orion, pronounced as though written O'Ryan. It is often called the "Wild Irishman of the Skies," and was always pictured by the ancients as a marvelous giant swordsman. This constellation may be located a little to the east and southward from Aldebaran. A line drawn from the Pole Star down through Capella and "produced as much farther again" strikes the western corner of a long irregular four-sided space in which may be seen the belt of the hunter, set as three slanting stars. Below these, as though dropped from them, are three more stars in the position of a sword. Two brilliant suns of the first magnitude mark this splendid figure unerringly for all eyes: Betelgeuse, with whose prowess we are already familiar, is located from its name, meaning "armpit"; Rigel, whom some astronomers term Algebar, is on the opposite lower corner. Bellatrix, "The Female Warrior or Amazon Star," is a star of the second magnitude, as are also the three stars in the belt and the star at the lower corner toward which they point. According to Porter, "The three stars in the belt constitute the golden yardarm of seamen, and the yardstick or ell of tradesmen, besides being popularly known as the Magi or three wise men from the Orient, and the three Marys." The same authority also tells us that by reason of the fact that Orion was supposedly swung up to heaven to teach men not to be too confident of their own strength, a committee of the University of Leipsic, in 1807, resolved that the stars belonging to the belt and sword of Orion should in the future be called Napoleon. Their resolution, however, failed to meet the approval of astronomers at large, and map-makers have given the suggestion small heed. As Orion lies on the celestial equator, he is equally familiar to the peoples of the North and the South, and he has been a favorite allusion with poets and writers of all time, being several times mentioned in the Bible. Dominating the skies as he does at a season when squally weather is rife, it is not to be wondered that he is blamed for much of the unpleasantness. Indeed an ancient writer goes so far as to state that the loss of the Roman fleet during the first Punic war was caused by the "obstinacy of the consuls, who, despite the pilots, would sail between the risings of Orion and Sirius, always a squally time."

Sirius, you remember, is the Dog Star, otherwise known as Canis Major, and the brightest star in all the heavens, being many times brighter than any other star of the first magnitude, and therefore in a class quite by himself. If the line of Orion's belt be drawn on downward in the direction indicated by these stars, Sirius will presently be met. But it takes a lively imagination to picture the figure of a dog in the five or six naked eye stars which make up this constellation! The name Sirius comes from the Greek, and signifies "scorching" or "sparkling." The ancients believed that the fierce rays from this star, when at its height, mingled with those of the Sun in the torrid days of summer, hence the term "dog days." Thus it transpired that, though marvelously lovely, Sirius soon attained a very unenviable reputation throughout the northern hemisphere, where it was believed his burning breath tainted the air with all manner of fevers, plagues, and death. In the Nile valley, however, Sirius proved himself quite as double in character, as he is in form; for here, before precession had carried him out of range, he rose with the Sun at the time of the summer solstice, and hence foretold the joyful coming of the rise of the waters. He was, therefore, hailed as the Nile star, the beloved "Star of Kneph," and many temples were dedicated to his worship. Unlike most doubles, Sirius shines with an intense white light. As he is always seen comparatively close to the horizon, his rays of light, which are more than eight years reaching us, are so broken up by their passage through the atmosphere, that he seems always to twinkle at a marvelous rate. This excessive scintillation gives the impression of a many-colored, changing light, which Tennyson thus describes in *The Princess:*

> The fiery Sirius alters hue,
> And bickers into red and emerald.

A line drawn slanting northeast from Sirius leads one to the constellation known as the Little Dog, or Canis Minor. This animal is a hound belonging to the famous "Heavenly Twins," at whose heels he follows. You will note these renowned hunters, glowing as two bright stars, a little to the westward, almost in line with Capella. Castor and Pollux they are called, and with the stars which make up their very sketchy outline form the constellation of Gemini, pronounced Jiminy. Henceforth, when you hear any one exclaim "By Jiminy!" bear in mind that it is by this constellation they are swearing! Castor and Pollux were the sons of Leda, and hence are known in literature as the Ledæan lights. In *The Wanderer,* Owen Meredith speaks of them thus:

> The lone Ledæan lights from yon enchanted air
> Look down upon my spirit, like a spirit's eyes that
> love me.

The twins were famous warriors, Castor being especially known for his miraculous battle on horseback, while Pollux was a pugilist to be feared. They were brothers of Helen of Troy, and on the return of the much recounted Argonautic expedition, when a storm threatened to destroy the vessel, Apollo, to whom an appeal was made, allayed the tempest, and set a star on the head of each Twin in token thereof. Henceforward The Twins became revered by seamen, and their effigies were frequently placed on boats as a talisman. The electrical flames, known as Saint Elmo's lights, which are frequently seen playing about the mastheads of the vessels in heavy weather, were formerly supposed to be due to the influence of these mystic

Twins. Thus in Longfellow's *Golden Legend* the padrone assures the prince:

> Last night I saw Saint Elmo's stars,
> With their glittering lanterns all at play
> On the tops of the masts and the tips of the stars,
> And I knew we should have foul weather to-day.

Castor does not appear as brilliant in the heavens as his brother, but he is in thruth more interesting. Mention has already been made of him as a famous double star, blue in color, and so gorgeous and stately, that centuries speed on while he revolves about his companion in regal majesty.

Returning again to the Great Bear, we draw a line through the last stars of his tail, continuing on until we come to Boötes, the Herdsman, whom modern peoples would, no doubt, style the "Ox-Driver," as he follows the plow oxen round and round the pole. The two bright stars in front of Boötes are the hounds which aided him in the chase. Between these two is a fine double star named Cor Caroli, which was discovered on the eve of the return of Charles II to London, and therefore christened in his honor. It has been recorded in history that this well-

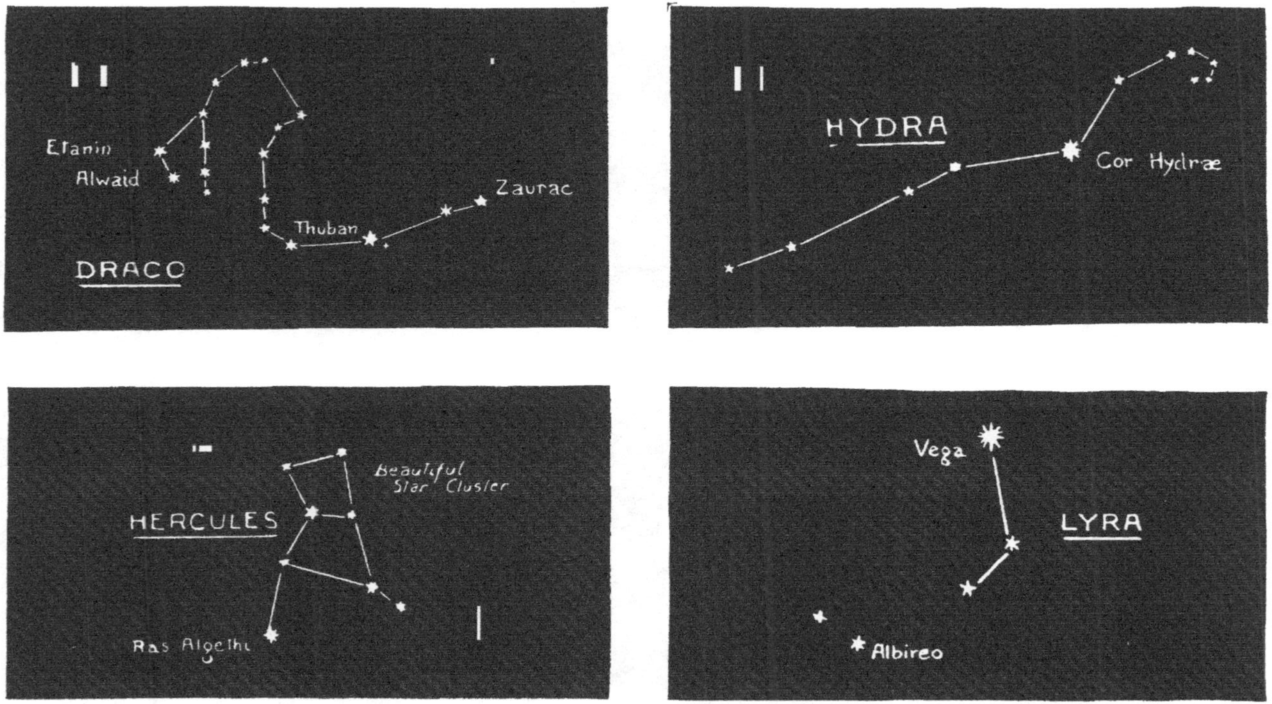

CONSTELLATIONS III

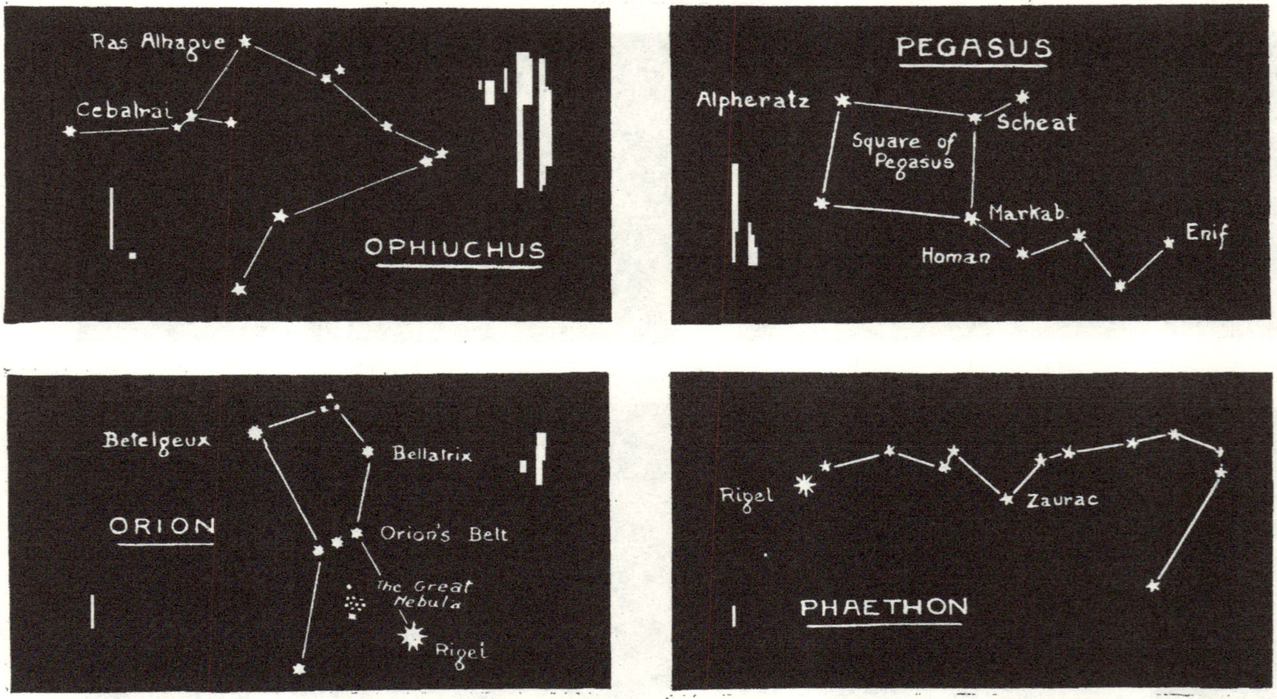

CONSTELLATIONS IV

known merry monarch "never said a foolish thing, and never did a wise one." But we must not forget that it was he who issued the decree for founding the royal observatory at Greenwich, thereby giving to the science of astronomy an impetus of incalculable value. In the early times, Arcturus, already mentioned as a wondrous star of the first magnitude, was represented as a spear in the hand of a hunter who was supposed to be pursuing the Great Bear. The Arabs termed it the "Lancebearer." Not far from Arcturus is a cluster of stars shaped like a horseshoe. This is the Northern Crown, or tiara of stars which the ancients connected with Ariadne, the daughter of King Minos, of Crete, who became the bride of Theseus while he was at the height of his Athenian triumphs. Later she was deserted by her faithless husband, and Venus, in pity, promised to send her an immortal lover. Accordingly Bacchus, the god of wine, wooed and won her, and gave her a most glorious crown of jewels as a wedding gift. On her death, this crown was transferred to the sky, and the gems became a wreath of sparkling stars. But the centuries seem to have dimmed their brightness. Today there is only one fairly bright star in the collection. However, the Northern Crown is interesting to look at, and it serves furthermore as a guide-post in locating the famous cluster of Hercules, which lies near at hand.

The wonderfully bright star which shines nearly overhead in summer is Vega, the harp-star, known also as the "arc-light of the sky," by reason of its brilliant sapphire hue. By looking closely you may be able to trace the outline of a harp among the small stars near. This represents the constellation of Lyra, supposed by the ancients to have been connected with the harp of Arion, a famous Sicilian musician who was once set upon by pirates, who sought to wrest from him a valuable prize which he had won by his skill. As they were about to throw him overboard, Arion requested permission to play for the last time on his harp. Nothing loath, the robbers settled themselves for a rare musical treat, and were not surprised when presently the dolphins began to gather about the boat in a charmed circle. Amused by the evident pleasure of the motionless herd, their eyes failed to take note of their prisoner, and ere they were aware he was mounted on the back of the largest dolphin and speeding off for the land near at hand. To-day this clever dolphin reposes among the stars, a little to the east of the harp. The ring nebula of Lyra is a favorite object for the ordinary telescope. Here, too, a little northeast of Vega, is a famous quadruple star, called the Double-double, whose duplicity may be noted by a keen eye. It takes a telescope, however, to note the two pairs. "Each pair," we are told, "seems to be revolving in one or two thousand years, and their common drift through space renders it probable that the two couples form one greater system, whose period of revolution must stretch far on towards a million years."

Southeast from Lyra are three bright stars in a row which are easily noticeable any starlit night from early summer to late autumn. They mark the constellation of Aquila, the Eagle. But clever indeed must be the imagination that traces the outline of an eagle in the surrounding stars! The ancients saw in it the ominous bird of Jove which bore his missiles of thunder. In its talons it carries the youth Ganymede, whom the Great Jove had caused to be seized and borne aloft to serve him as a cup-bearer.

> . . . flushed Ganymede, his rosy thigh
> Half buried in the eagle's down
> Sole as a flying star shot through the sky
> Above the pillared town.
>
> —*Tennyson.*

The Little Bear and Lyra form the guide-posts for another constellation that is associated with a well-known tale. We remember that, when Phaëton swept from his ill-starred voyage as charioteer of the Sun, his friend Cygnus could not give him up, and dived so often in the stream in search of his body that the gods grew angry and changed him into a swan. Later the pensive bird, which still sailed hither and yon, thrusting its head into the water, was caught up into the sky, where it forms the figure known as Cygnus, the Swan. We will do marvelously well, however, if we succeed in tracing an outline which in any way resembles this water fowl! But the principal stars of this constellation form the Northern Cross, and this you may readily recognize. It is well known to all stargazers. Lowell gives a beautiful picture of the Cross presiding over the opening of the New Year:

> Orion kneeling in his starry niche,
> The Lyre whose strings give music audible
> To holy ears, and countless splendors more,
> Crowned by the blazing Cross high-hung o'er all.

Albireo, in the Swan's head, is a lovely double star, formed by two companions of golden and azure, so well separated that they make exquisite objects in the small telescope. 61 Cygni, in the left wing, has been mentioned variously in these pages.

The Southern Cross, though in no wise connected with the constellation of the Swan and invisible within the boundaries of the United States may be mentioned just here. It forms "the night-clock" of those who live within and beyond the tropics. A writer of the southlands remarks: "How often have we heard our guides exclaim in the savannahs of Venezuela and in the desert extending from Lima to Truxillo, 'Midnight is past, the cross begins to bend!'" Whittier in his *Cry of a Lost Soul,* cites the Southern Cross as a symbol of God's mercy. He pictures a traveling party in the gloomy forests of the Amazon, when suddenly through the night lists,

> A cry, as of the pained heart of the wood,
> The long, despairing moan of solitude.

The guides cross themselves and explain in low frightened voices that it is the plaint of a lost soul, some unbeliever burning in hell. But there is one, who unconvinced,

> Lifts to the starry calm of heaven his eyes,
> And lo! rebuking all earth's ominous cries,
> The Cross of pardon lights the tropic skies.

THE ZODIAC

Familiarity with the constellations already mentioned will give the amateur a wide range of acquaintances in the sky, and will perhaps be sufficient for those who do not wish to delve more deeply into astronomy. For the others, however, there remains yet an interesting trail. It is the path marked by the zodiac among the stars. The zodiac, you remember, is an imaginary zone or girdle, stretching around the celestial sphere, in a belt extending eight degrees on either side of the ecliptic. Neither the Moon nor any one of the planets can ever travel outside this belt. It forms what has been termed "the zoölogical garden of the sky," since it is divided into twelve animal groups, and the term *zodiac* itself signifies the Greek for "a circle of animals." How ancient this division is cannot be told, but it is certain that it has at least come down from pre-Babylonian peoples. To this zodiacal division is traceable, no doubt, the twelve months which make up the solar year, each sign measuring the progress of the Sun during one complete revolution of the Moon. Originally the division of the zodiac into constellations served as the best possible means of noting the positions of the heavenly bodies. But since the "animals" named are more or less imaginary and rambling, extending as each one does over thirty degrees of space, accurate astronomy of to-day discards this scheme, and positions are taken with reference to the ecliptic and their celestial longitude. We still speak, however, of the Sun and the planets as being in certain constellations, and they are so indicated in our calendars and almanacs. Let us, then, try to understand what is meant.

Imagine the Earth moving round its orbit with the Sun in the middle. Now around this draw another wide ellipse to represent the zodiac and divide it into twelve equal parts, which you understand remain always fixed. As the Earth moves, a person located at any one point will apparently see the Sun continually against a different background—that is to say he seems to change his position among the stars and to move in and across the constellations by reason of our revolution. The same rule holds true of the Moon and the planets. Thus we say that a heavenly body is *in* a certain constellation when it seems for the time being to be a part of it. As the planets are always to be found in some one of the zodiac constellations, and as the almanacs tell us which one this is, we have only to learn the stars of the zodiac to locate any planet we wish to plot. A sure aid in locating the zodiacal constellations is to know at what time they will be on our meridian, that is the line over our head due north and south, at nine o'clock at night, during a given month. Here is a table by months.

Astronomical Name	Common Name	Time
Aries	The Ram	December
Taurus	The Bull	January
Gemini	The Twins	February
Cancer	The Crab	March
Leo	The Lion	April
Virgo	The Virgin	May
Libra	The Balance	June
Scorpio	The Scorpion	July
Sagittarius	The Archer	August
Capricornus	The Goat	September
Aquarius	The Water-Bearer	October
Pisces	The Fishes	November

To find any zodiacal constellation during any other month than that here given, subtract two hours for each following month. Suppose we want to find Aquarius, the water-bearer, in November instead of October. We must look for it on our meridian at 7 P. M., and in December, at 5 P. M.

Owing to the precession or backward movement of the equinoxes, a confusion has resulted in the *constellations* of the zodiac and the *signs* of the zodiac. Time was when they were one and the same; now the two terms stand for something different. The signs of the zodiac are symbols for the spaces or parts of the great circle which it forms. Two or three thousand years ago, when they were christened, the constellations stood in the signs of their name; since then, these signs have shifted one space to the right, while the constellations have, of course, stood still. The result is a tangle which, however, can be unraveled after a few efforts. When the almanac says that a planet is in that part of the zodiac whose sign is Aries, this does not mean that it is in the constellation of Aries. For the constellation of Aries has apparently moved westward and is now in that part of the zodiac indicated by the symbol Taurus; the constellation of Pisces occupies the place indicated by the sign of Aries. We shall have no difficulty with this if we remember always to shift the constellation one place to the right of the sign which bears its name.

We must not expect to locate the stars of the zodiac by the constellations as outlined in the diagrams of the Circle of the Zodiac. Remember that for the purposes of illustration these had to be made compact; in reality the stars of these figures are scattered widely over thirty degrees of space, as has been mentioned, and some do not even stop there, but range on to some particular constellation which stands as a guide-post. We

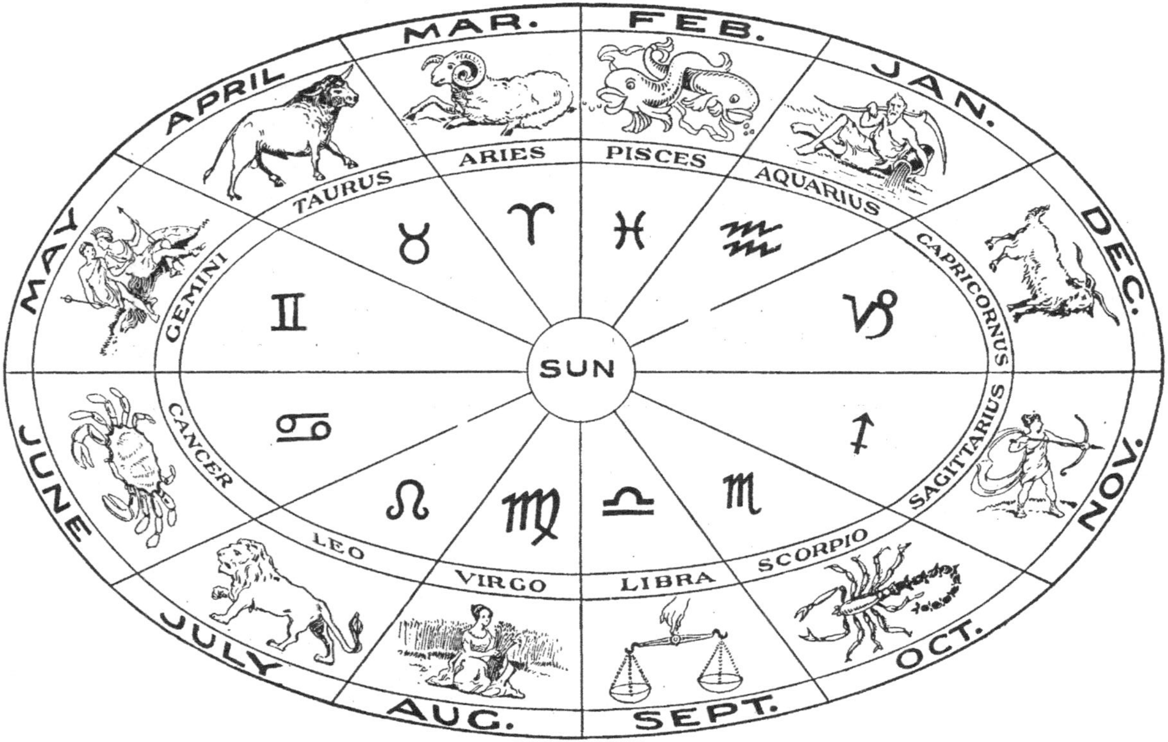

SIGNS OF THE ZODIAC

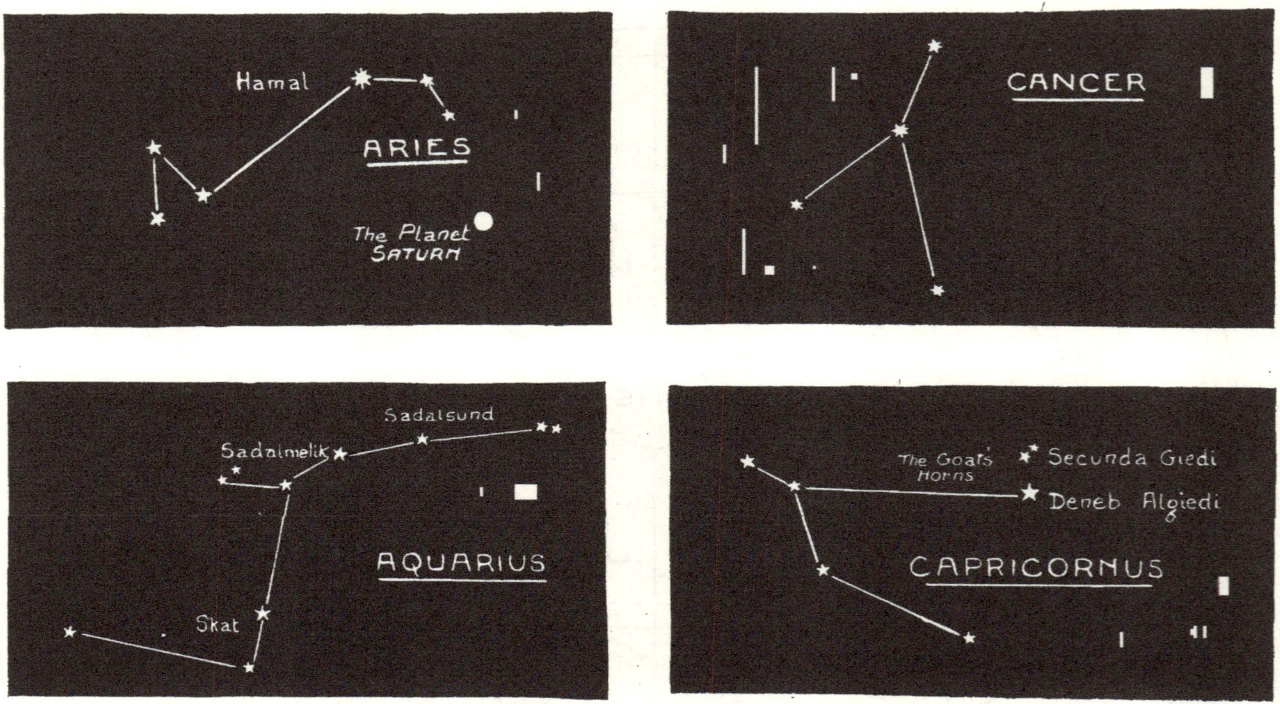

CONSTELLATIONS OF THE ZODIAC I

will need a good star map to make any real headway in locating the stars of the zodiac. In the almanacs our year begins with January. The sun is then in the sign Aquarius, hence this month is usually indicated by the picture of Aquarius pouring water from a pitcher. In studying the zodiac, we begin with Aries, which the sun formerly entered in March, and read from right to left or counter-clockwise, in the position which the Earth travels, thus following the apparent eastward path of the sun through the constellations. The following rhyme puts the order of the stars of the zodiac firmly in mind:

The Ram, the Bull, the Heavenly Twins,
And next the Crab, the Lion shines,
The Virgin and the Scales;
The Scorpion, Archer, and He-goat,
The Man that holds the watering-pot,
The Fish with glittering tails.

With some of these constellations we are already familiar. It gives us a little heart for the task,—a beginning in the great elongated wheel or circle which outlines the ecliptic. Quickly we search out the Bull and the Heavenly Twins. But the Ram heads the list: how shall we find him? Easily enough, as it happens. We have only to draw a line from the Pole Star to Alpha in the lower left-hand corner of the Great Square of Pegasus; then downward until we come to two bright stars quite close together, and a third one not quite so bright. These are the chief stars in the constellation of Aries, the Ram with the wondrous golden fleece of Argonautic fame.

Cancer, the Crab, is the most difficult to locate of all the zodiacal constellations. He stands at the sign which marks the northern tropic, where the Sun stops ascending. According to old Chaldean philosophers it was just here that "the gate of men" was located, by which human souls descended to fill the bodies allotted to them. Capricorn, in the southern tropic, where the Sun again turned, was the gate where the souls were received back to heaven. In the constellation of Cancer is the famous beehive cluster, already mentioned. Mythology styled this the Manger, the two stars lying one on either side being named the ass's colts. The Manger has not infrequently been mistaken for a comet by inexperienced eyes. An old writer makes it the basis for a clever weather prognostication:

A murky manger with both stars
Shining unaltered, is a sign of rain.
If while the northern ass is dimmed
By vaporous shroud, he of the south gleam radiant,
Expect a south wind: the vaporous shroud and radiance
Exchanging stars, harbinger Boreas.

Two lines extending down from either side of the Dipper, will reach the two chief parts of Leo, the Lion, the head and shoulders, or what seems to us a sickle, composed of six stars, with Regulus, a very bright sun, meaning "little king," in the end of the handle. Needless to say, it takes a good deal of imagination to picture the king of beasts in such a scraggly outline! Much less can we fancy it as the leader of the four royal guardians of heaven, or account for its wide-spread prestige of greatness and power, which brought glory and riches to all luck enough to be born under its magic influence.

By drawing a line from the Pole star through Mizar in the handle of the Big Dipper, on down past Boötes, until a big white star is reached, we arrive at Spica, the Virgin, the chief star in the constellation of Virgo. This group is of interest chiefly because while the Sun is at this point it passes the autumnal equinox. The Virgin and the harvest were always inseparably linked in the minds of the ancients, and she is invariably represented with a sheaf of wheat in one hand and a sickle in the other. Instead of the goddess of the fields, modern minds see in this group a huge star diamond outlined against the blue. It is the largest of the zodiacal constellations, and stands in the field where there are comparatively few stars, being quite remote from the Milky Way, in a region where the nebulæ congregate in rich numbers.

The scales mark the constellation of Libra, the Balance. "In all the round of the zodiac," says Porter, "this constellation alone represents an inanimate object; and its antiquity, though somewhat in dispute, does not seem to be very great. Certain it is that the Greeks associated its stars with the claws of the Scorpion which follows to the east. Scorpio, indeed, seems to have been considered a double sign, thus completing the twelve."

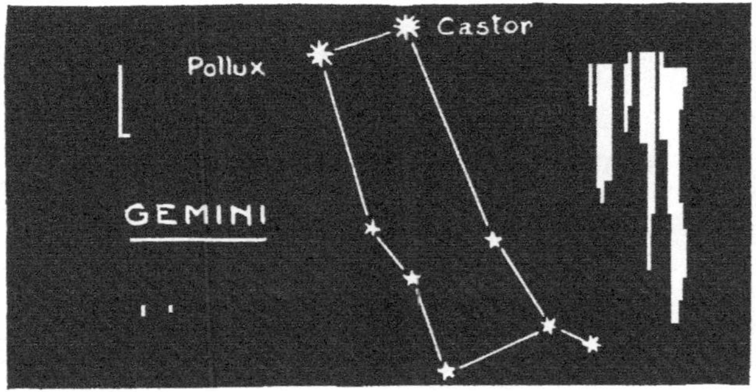
Castor
Pollux
GEMINI

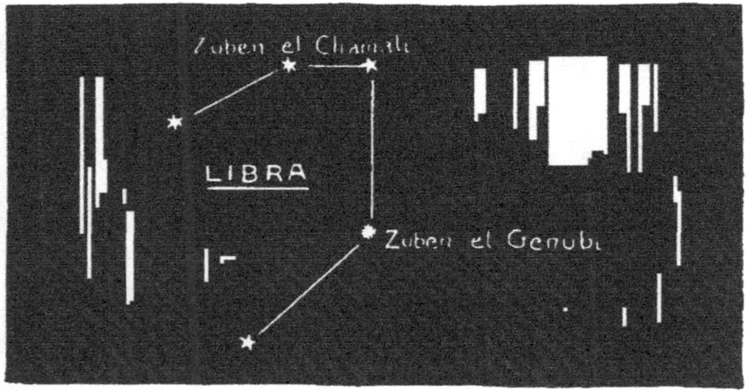
Zuben el Chamali
LIBRA
Zuben el Genubi

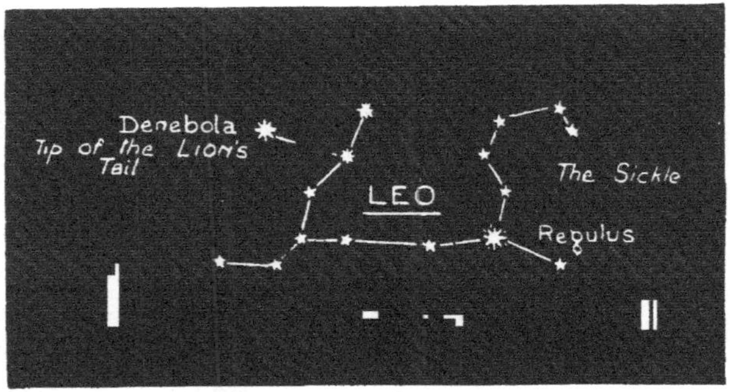
Denebola
Tip of the Lion's Tail
LEO
The Sickle
Regulus

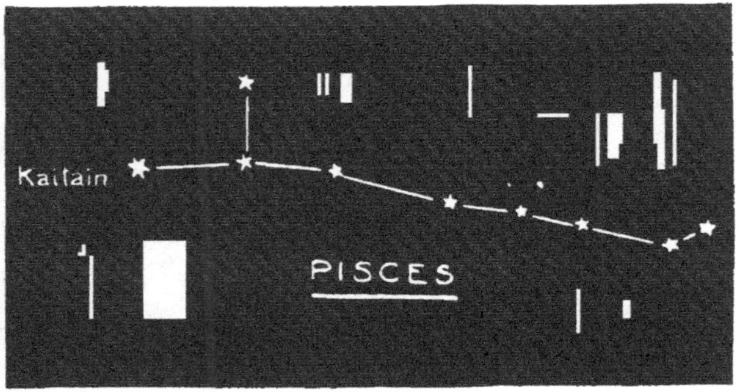
Kaitain
PISCES

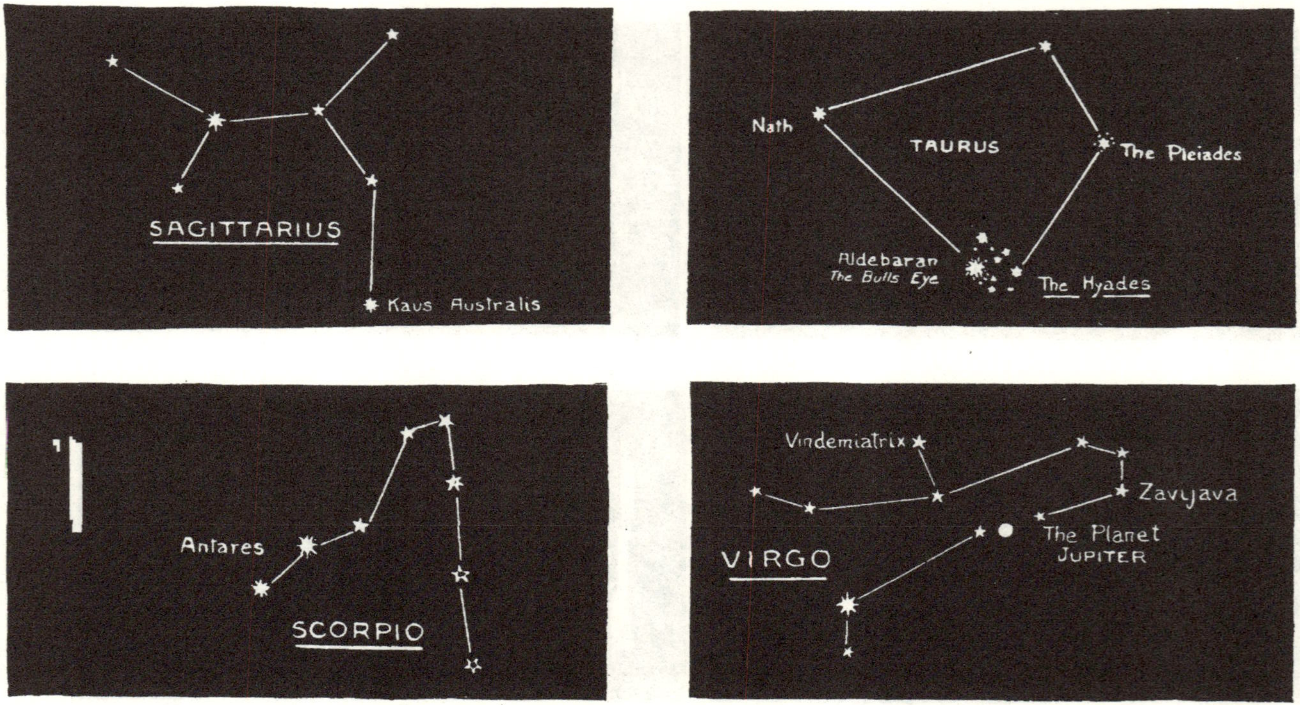

CONSTELLATIONS OF THE ZODIAC III

I bear the scales, when hang in equipose
The night and day.

Thus wrote Longfellow, in his *Poet's Calendar,* striking the key note for the birth of this constellation, which took place at a period when the autumual equinox was in the sign of the zodiac, thereby suggesting the figure of *the balance,* corresponding to equal nights and days. The constellation is a rather dim and poor affair, and we should probably miss it altogether save that Alpha, its brightest star, is the connecting link between Spica in Virgo and Antares in Scorpion. The latter constellation is one of the few which really looks its part. Its heart is the big bright red star Antares, the king of suns, and an interesting double in the bargain, whose green companion lies so near as to be frequently enshrouded in his flaming rays. The name Antares comes from the Greek and signifies "the rival of Mars," this star being in truth the only one in the heavens which could by any chance be mistaken for the red planet. When Mars chances to be in Scorpion it takes nice discrimination to tell them apart.

Sagittarius, the archer, is christened in honor

of the ancient centaur or man-horse. He is the patron of the hunt and the chase, and is always pictured shooting an arrow from a bow at the fiery heart of the Scorpion. We find Sagittarius easy to locate, as it lies where the path of the zodiac crosses the Milky Way. One of its most outstanding features is the little milk dipper, formed of seven stars, and lying turned upside down. There are also several interesting star clusters and nebulæ in this constellation. The winter solstice is marked half-way between the constellation of Scorpion and Sagittarius, at a point between the two streams of the Milky Way. In ancient times Capricornus, the sea-goat, marked this solstice; then the constellation occupied the lowest or most southern part of the zodiac. Note Milton's lines:

> Some say the Sun
> Was bid turn reins from the equinoctial road
> . . . .
> Up to the tropic Crab; thence down amain,
> By Leo and the Virgin and the Scales,
> As deep as Capicorn, to bring in change
> Of seasons to each clime.

Capricornus is now a constellation of autumn. Of course, he looks no more like a goat-fish than

Sagittarius looks like a man-horse. But so the early peoples pictured him, and the astrologers reckoned that person fortunate indeed who was born under this sign of the zodiac. Capricornus is easily found by drawing a line through Vega in Lyra and on to Altair in Aquila and thence to the path of the zodiac. The two brightest stars of Capricornus are in his goat-like head. One of these stars is a famous naked-eye double. In the days of the ancients, however, it took an especially good eye to detect this, and hence it belonged to the "test" stars. The doubles, we are told, "are separating at the rate of one minute of an arc in about thirteen hundred years, their present distance being six minutes." This corresponds to a distance of about forty-eight quadrillion miles. Another bright star of the sea-goat is to be found in his fish-like tail. Of course, it was inevitable that Capricornus should become associated in mythology with the god Pan, who was also part goat. Indeed, Capricornus is himself the god Pan in his goat-like disguise; for, being attacked by a great fire-breathing typhon, Pan jumped into the sea to save himself. Here he shortly threw out a finny tail and became amphibious. Pictures of this constellation, therefore, always show a curious animal with the head and horns of a goat and a body which slopes off into the hinder parts of a great fish.

Aquarius is the man with the watering-pot. His is a constellation of autumn, but the stars which form it are faint and hard to find. It is located in that part of the sky which the Chaldeans designated as "the sea." The dolphin, Cetus the whale, three fishes, and a goodly company of other aquatic creatures are among his near neighbors. The Romans gave to him his name of *the Waterman,* because in the long ago when the sun entered the sign of Aquarius, in January, there were usually heavy rains throughout Italy.

The "Fish with glittering tails" is the constellation of Pisces. This last member of the zodiacal group is really twins, being fashioned of two fishes, widely separated, but having their tales connected by a ribbon. It is not an interesting constellation to look at with the naked eye, for its stars stream out in two faint lines, spreading away beyond their own division of the zodiac in a confusing fashion that is hard to follow. But Pisces is none the less a very important constellation, because it is at one of the points where the line of the equator crosses the ecliptic. This occurs about the twenty-first of March, or at the time of the spring equinox, which, because the days and the nights are then of equal length all over the earth, is called the *vernal equinox.* Pisces is very nearly starless. Some ancient peoples referred to this constellation as "Venus and Cupid," the mythological tale being that Venus, having been frightened by the giant Typhon, threw herself and the child Cupid into the Euphrates to escape from him. The gods, in pity, changed them to fishes, and later fixed them in the sky. Formalhaut, the bright star in the constellation known as the "Southern Fish," is located south of Aquarius and should not be confused with the fishes of Pisces.

"THE INFINITE MEADOWS OF HEAVEN"

Outside the familiar constellations and the more complicated path of the zodiac, there yet remains a limitless host for those who would go farther. One thing deters: As we consider the universe of the stars, we find ourselves utterly without adequate scales for measurement. Even the vast scale of millions of miles used in determining the points of the solar system is useless here. We must think in billions and hundreds of billions of miles. We are lost in a marvel of Infinity and Eternity—a Universe totally without bounds, having absolutely no beginning or end in time. And thus it must ever be:

I open my scuttle at night and see the far-sprinkled systems,
And all I see, multiplied as high as I can cipher, edge but the rim of the farther systems:
Wider and wider they spread, expanding, always expanding,
Outward, outward, and forever outward:
My sun has his sun, and around him obediently wheels;
He joins with his partners a group of superior circuit,
And greater sets follow, making specks of the greatest inside them.

"There is no stoppage, and never can be stoppage.
If I, you, the worlds, all beneath or upon their surfaces, and all the palpable life, were this moment reduced back to a pallid float, it would not avail in the long run.
We should surely bring up again where we now stand,
And as surely go as much farther—and then farther and farther.

A few quadrillions of eras, a few octillions of cubic leagues,
do not hazard the span or make it impatient.
They are but parts—anything is but a part,
See ever so far, there is limitless space outside of that,
Count ever so much, there is limitless time around that.

—Walt Whitman.

PRINCIPAL STARS AND PLANETS

Name in parenthesis is constellation to which the star belongs.

ACHERNAR *(Eridanus)*. A star of the first magnitude, eighth in order of brightness. Invisible in northern latitudes. Crosses meridian December 16th. Receding at rate of ten miles per second.

ALBIREO *(Cygnus)*. A third dimension star in opposite extremity of constellation from Arided. Actually a double star.

ALCOR *(Ursa Major)*. Companion star to Mizar.

ALDEBARAN *(Taurus)*. A star of the first magnitude, noted for its reddish tint. Its luminosity is about twenty-three times that of our Sun. Distant twenty-nine light years. Known as "the Bull's Eye." Receding at rate of thirty miles per second. Distant thirty-two light years.

ALGOL *(Perseus)*. A famous variable star which waxes and wanes, decreasing in magnitude from 2.4 to 3.6. Algol forms part of the head of Medusa, which Perseus is supposed to carry.

ALKAID *(Ursa Major)*. Epsilon, or fifth star in brightness in the "Dipper." Of the second magnitude. End of handle in "Dipper."

ALPHA ANDROMEDÆ *(Andromeda)*. A double star of bluish-green tint and great brilliance. 2.5 to 5.5 magnitude.

ALPHA CASSIOPEIÆ *(Cassiopeia)*. A twin star, the

brightest in this northern constellation, which is actually a pair of blazing suns. Each is nearly the same size as our Sun.

ALPHA CENTAURI *(Centaurus).* A fine double star of the first magnitude, one being almost the counterpart of our Sun. Ranks third in order of brightness. Invisible in northern latitudes. Crosses meridian June 29th. Approaching at rate of 13.8 miles per second. Distant four light years.

ALPHA CRUCIS *(Crux).* The brightest star in the Southern Cross, but not visible in the northern latitudes. It is a triple star of 1.5, 1.8, and 6 magnitude. Crosses meridian May 29th. Receding at rate of 4.3 miles per second.

ALPHA PERSEII *(Perseus).* Brightest star in this eastern constellation. Set in a particularly fine star field.

ALPHECCA *(Corona).* Brightest star in this constellation.

ALTAIR *(Aquila).* A brilliant star whose light is about eight times that of the Sun. Of first magnitude, and thirteenth in order of brightness. Crosses meridian September 19th. Approaching at rate of 20.5 miles per second. Distant sixteen light years.

ANDROMEDA NEBULA *(Andromeda).* A nebular cluster of unusual beauty, and the brightest nebula in the heavens.

ANTARES *(Scorpius).* A blazing red star about 3,000 times as bright as our Sun. Of the first magnitude. Said to be even larger than Betelgeuse.

ARCTURUS *(Boötes).* The brightest star in northern sky. Its light is 500 times that of the Sun; its velocity about 200 miles a second. Ranks fifth in order of brightness. Crosses meridian June 24th. Distant 160 light years.

BELLATRIX *(Orion)*. A star of second magnitude in opposition to Betelgeuse.

BETA CASSIOPEIÆ *(Cassiopeia)*. The right "pointer" at top of "W," which is used in determining Greenwich time.

BETA CENTAURI *(Centaurus)*. The tenth star in order of brightness, but invisible in northern latitudes. Crosses meridian June 21st.

BETA CYGNI *(Cygnus)*. An extremely beautiful double star.

BETELGEUSE *(Orion)*. The brightest star in this constellation, and one of the largest known. Its mass is now reckoned to be 43 million times that of our Sun. Of the first magnitude, and eleventh in order of brightness. Crosses meridian February 15th. Receding from the Earth at rate of thirteen miles per second.

CANOPUS *(Argus)*. Outranked only by Sirius in order of brightness. A first magnitude star, but invisible in our middle northern latitudes. Crosses meridian February 23rd. Receding from the Earth at rate of 12.7 miles per second.

CAPELLA *(Auriga)*. One of the brightest stars, and of first magnitude. Its mass is about 4000 times that of our Sun, and its luminosity 130 times as great. Ranks fourth in order of brightness. Crosses meridian February 5th. Receding from Earth at rate of 19.7 miles per second. Distant thirty-two light years. Near Capella may be seen the "Three Kids."

CASTOR *(Gemini)*. A very fine double star. Magnitude 2 and 2.8. Forms with Pollux the "Heavenly Twins."

CERES. A minor planet whose diameter is 480 miles.

Cor Caroli *(Boötes)*. A fine double star named in honor of Charles II.

61 Cygni *(Cygnus)*. Famous as the first star whose distance was ever measured. Actually a double star, of 5.6 magnitude. Velocity sixty-three miles per second. Distant seven light years.

Deneb *(Cygnus)*. A star of the first magnitude, and about twentieth in order of brilliance.

Dubhe *(Ursa Major)*. Alpha, or brightest star in the "Dipper." Of the second magnitude. Dubhe and Merak are the "pointer" to the North Star, Dubhe being the nearer.

Earth. Planet of our solar system. Third nearest the Sun. Distant 93 million miles. Yearly orbit 365 days. Diameter 7913 miles.

Eros. A minor planet with an elliptical orbit.

Formalhaut *(Piscis Australis)*. A star of the first magnitude, ranking eighteenth in order of brilliance.

Hamal *(Aries)*. Brightest star in this constellation. Of second magnitude.

Jupiter. Planet of our solar system. Distant 483 million miles from the Sun. Time of orbit twelve years. Size 1200 times that of the Earth. Diameter 92,000 miles.

Markab *(Pegasus)*. One of four corner markers of this square. A second magnitude star.

Mars. Planet of our solar system. Fourth nearest the Sun. Distant 141 million miles. Time of orbit 687 days. Diameter 4230 miles.

Merak *(Ursa Major)*. Beta, or second brightest star in the "Dipper." Of the second magnitude. Forms with Dubhe the "Pointer" to the North Star, Dubhe being the nearer.

Mercury. Planet of our solar system. Nearest the Sun.

Distant 36 million miles. Time of orbit eighty-eight days. Diameter 3030 miles.

MESSIER 13 *(Hercules)*. A fine star cluster which has been resolved by powerful telescopes into a group of 5000 stars, some of them comparable to our Sun.

MIZAR *(Ursa Major)*. A second magnitude star at the bend of the handle of the "Dipper." Near it may be seen a companion star, Alcor, of the fifth magnitude. Mizar itself is a twin star, radiating 115 times as much light as the Sun. Distant seventy-five light years. The two stars are 25 million miles apart.

NEKKAR *(Boötes)*. A third magnitude star in the apex of this constellation, furthest removed from Arcturus.

NEPTUNE. Planet on extreme outer edge of our solar system. Distant 2790 million miles from the Sun. Time of orbit 165 years. Diameter, 34,000 miles.

ORION NEBULA *(Orion)*. One of the finest nebular masses in the heavens. Appears as a hazy star to the unaided eye.

PHAD *(Ursa Major)*. Gamma, or third brightest star in the "Dipper." Of the second magnitude.

PLEIADES *(Taurus)*. A well-known cluster of bright stars, the central one being Alcyone. This group consists of seven stars, six being of the fourth magnitude; besides many of lesser brightness. The cluster crosses the meridian November 17th. The largest stars are 100 to 200 times as bright as the Sun.

POLARIS *(Ursa Minor)*. The Pole Star which gives us the due North of the heavens. Although not of the first magnitude, it is the most important of all, for this reason. Distant 47 light years. Around it all other stars and constellations appear to revolve.

POLLUX *(Gemini)*. A star of the first magnitude. Companion to Castor, the two being known as the "Heavenly Twins." Receding at rate of 2.4 miles per second.

PROCYON *(Canis Minor)*. One of our nearest star neighbors, and of the first magnitude. Its light is about ten times that of the Sun, its velocity eleven miles a second. Ranks ninth in order of brightness. Crosses meridian March 14th. Approaching at rate of 2.5 miles per second. Distant twelve light years.

REGULUS *(Leo)*. A bright star of 1.2 magnitude. Its light is 1,000 times greater than our Sun. Approaching at rate of five miles per second.

RIGEL *(Orion)*. A splendid star of the first magnitude which is 2,000 times as luminous as the Sun. Ranks seventh among the stars for brightness. Crosses meridian February 5th. Receding at rate of 13.6 miles per second. On opposite side of constellation from Betelgeuse.

SATURN. Planet of our solar system. Distant 886 million miles from the Sun. Time of orbit, twenty-nine and one-half years. Diameter, 74,000 miles. Noted for its rings.

SCHEAT *(Pegasus)*. One of four corner markers of this square. Of second magnitude.

SIRIUS *(Canis Major)*. The brightest star in the heavens. Magnitude 1.4. Although Sirius is only two and one-half times the mass of the Sun, it is thirty times as luminous. Crosses meridian February 28th. Approaching at rate of 5.6 miles per second. Distant 8.4 light miles.

SPICA *(Virgo)*. A double star of the first magnitude. The components are about 11 million miles apart, and are about nine and six times the mass of the Sun respectively. Receding at rate of 1.2 miles per second.

URANUS. Planet of our solar system, next to the farthest

removed. Distant 1780 million miles from the Sun. Time of orbit eighty-four years. Diameter, 31,000 miles.

VEGA *(Lyra)*. A star of first magnitude, and sixth in order of brilliance. About 100 times as radiant as our Sun. Crosses meridian August 30th. Approaching at rate of 8.5 miles per second. Astronomers reckon that it will become the North Star, 12,000 years hence. Distant twenty-seven light years.

VENUS. Planet of our solar system. Second nearest to the Sun. Distant 67 million miles. Time of orbit 225 days. Diameter 7799 miles, or nearlv the size of the Earth.

GLOSSARY

Aerolite. A mass or fragment falling from celestial space to the earth. During its flight it is called a meteor, and its fragments are meteorites or aerolites.

Aphelion. The point in space on the orbit of a planet or comet, when it is the furthest removed from the Sun. Reverse of *perihelion.*

Arc. A portion of the path of a heavenly body. A curve or part of a circle.

Asteroids. A group of small planets, or planetoids, revolving around the Sun. There are several hundred of these, and their diameters are usually less than 100 miles each.

Astronomy. That science which treats of the heavenly bodies, their size, orbit, distance, and relation to each other.

Aurora Borealis. The northern lights, or the glow which is often seen in the sky at the far north. Now believed to be electrical in origin.

Axial motion. That motion of a planet or other body around its axis, or common center.

Axis. An imaginary line through the center of a body, around which it revolves. The Earth's axis passes through its center from Pole to Pole.

Binary Stars. A pair of stars revolving around a common center of gravity.

Bolide. A shooting star or meteor, of unusual brilliance.

Centrifugal Force. A force pulling away from the

center of a body. The reaction or pull against a force that is causing a body to move in a circle.

CHROMOSPHERE. A sphere or layer of red gases surrounding the Sun. Visible during an eclipse.

COMA. The nebulous mass surrounding the head or nucleus of a comet.

COMET. A heavenly body noted for its singular appearance and eccentric orbit. It usually consists of a bright star-like nucleus, with a train of great length.

CONJUNCTION. The nearest point of approach of two heavenly bodies to each other, as viewed from the Earth.

CONSTELLATION. A fanciful grouping of certain neighboring stars to form a figure or picture. As the constellation of *Ursa Major* comprises 7 stars said to resemble a bear or dipper.

CONSTELLATIONS OF THE ZODIAC. The constellations or signs of the zodiac are 12 different groups of stars, one for each month, as determined originally by the fact that such constellation was in the zenith on a given month.

COPERNICAN THEORY. The accepted astronomical system, first advanced by Copernicus in 1543, which makes the Sun the center around which the Earth and other planets revolve.

CORONA. The circle of light radiating from a celestial body. Specifically, the Sun's corona is the mass of rays seen shooting from it in every direction, during an eclipse.

CRESCENT MOON. The visible portion of the Moon in its first or last quarter. Consequently, the term to denote a new moon, or one in its last days.

CREPE RING. A semi-transparent ring around Saturn—one of three rings, the other two being bright.

DEGREE. A 360th part of a circle or circumference. A ninetieth part of a right angle. Thus from due north to due east is a rectangle, and one degree is a 90th part of that distance.

DISPLACEMENT. The apparent change of position of a heavenly body from its true course.

DOUBLE STAR. Two stars so closely associated as to appear to be one, to the naked eye.

EARTH-SHINE. The faint light visible on the dark part of the Moon when it is new, and caused by reflection of light from the Earth.

EASTWARD. Running in an easterly direction, or at a right angles to the line of the North Star.

ECLIPSE. The obscuring of one heavenly body by another in the same line of vision—as the eclipse of the Sun by the passage of the Moon across its field.

ECLIPTIC. That plane, passing through the center of the Sun, which contains the Earth's orbit.

ELLIPSE. The path of a planet around the Sun, which may be popularly described as an oval or a flattened circle.

EQUATOR. The imaginary line passing around a planet at its widest part, and equidistant from its Poles.

EQUINOX. One of two opposite points in the heavens where the Sun's position in regard to the Earth causes days and nights of equal length.

EVENING STAR. The bright planet visible in the West just after sunset, and before any stars are visible.

FACULAE. Small spots on the Sun which are brighter than the rest of the photosphere.

FIRST QUARTER. The first of the four phases through which the Moon passes each month, and in which it appears as a crescent, then a quarter-filled circle.

FIXED STAR. A celestial body shining with its own light,

and so far removed from our own solar system that its relative position seems unchanged.

Galaxy. A luminous band of star clusters. Usually applied to the Milky Way.

Gegenschein. A faint luminous spot which may be seen on the ecliptic ninety degrees away from the Sun. Also called the zodiacal afterglow.

Gibbous. Rounded or swelling. A gibbous moon is one that is between half full and full.

Gravitation. The constant force or pull exerted between two bodies in space; also the tendency of all loose particles on any heavenly body to fall toward its center.

Grimaldi. A large plain on the surface of the Moon, said to contain 14,000 square miles.

Harvest Moon. The full moon occurring about the time of the autumnal equinox, which rises about the same time for several nights in succession.

Heliometer. An instrument used in measuring angles and distances between heavenly bodies. A specially constructed telescope whose objective may be cut into two parts.

Horizon. The point of one's vision where the earth and sky meet. The furthest point in any direction by land or sea, where the sky cuts off the view.

Hunter's Moon. Name given to the full Moon occurring in October.

Inclination of the Earth's axis. A leaning away from the upright, or from a right angle. The dip away from an upright position in the Earth's orbit, causing its axis to point to the north star.

Inner Planets. Name given to the four planets which are nearest the Sun—Mercury, Venus, Earth, and Mars.

INTERNATIONAL DATE LINE. The 180th meridian, or the one on the opposite side of the globe from the Greenwich Observatory, crossing the Pacific Ocean. Here time is set forward or back one day, according as one is traveling West or East.

LAST QUARTER. The fourth of the Moon's phases, when it is again approaching crescent shape and reaching "the dark of the Moon."

LIGHT PRESSURE. A constant repellent force which is exerted by the Sun's rays, but is felt only upon the lightest bodies, such as the tail of a comet, causing it always to point away from the Sun.

LIGHT-SPHERE. The radiant envelope of gases surrounding the Sun.

LIGHT YEAR. A term used by astronomers to denote extreme distances. Light travels 180,000 miles per second, and in one year would travel nearly seven and one-half billion miles.

MAGNITUDE. An arbitrary scale for measuring the brightness of stars, the brightest (twenty in number) being of the *first magnitude,* and the faintest, as seen by the unaided eye, the *sixth.*

MEAN SOLAR DAY. The average length of the total number of days as seen in a transit around the Sun. The common measure of time.

MERIDIAN. The line in the heavens running north and south and passing directly overhead. The vertical plane parallel to the Earth's axis.

METEOR. Name given to a small heavenly body flying through space which suddenly strikes the Earth's atmosphere and becomes luminous.

METEORITE. A mass of stone, iron, or other metal which has fallen to the Earth from the outer realms of space.

MIDNIGHT. The middle of the night—twelve o'clock—

when the Sun is on the opposite side of the Earth.

MILE. The unit of measure for long distances in English-speaking countries. One-sixtieth of a degree of the Earth's surface at the equator. In America, 5280 feet.

MILKY WAY. Also called the Galaxy, the Milky Way is the broad band or bands of hazy star clusters and nebulæ stretching across the heavens.

MINUTE. In time, the sixtieth part of an hour. In astronomy, the sixtieth part of a degree.

MOON. A celestial body revolving around the Earth once in about 27 days, 8 hrs. Its satellite. The name, *moon,* is also applied to the satellites of other planets.

MORNING STAR. The planet which is conspicuous in the eastern sky just before dawn. Jupiter, Mars, Saturn, or Venus.

NEAP TIDES. The tides occurring just after the first and third quarters of the Moon, when its attraction upon the sea is weakest.

NEBULA. In astronomy, an unformed mass of stellar substance, which is believed to be stars and planets in the making.

NEBULAR HYPOTHESIS. The theory advocated by Kant, Herschel and others, that our solar system was originally a nebula which as it cooled and contracted formed the Sun and its planets.

NEW MOON. The first phase of the Moon, in which it appears as a thin, bright crescent.

NODE. The point where the orbit of a heavenly body cuts across the ecliptic. The intersection on the celestial sphere of any two circles, such as the equator and the ecliptic.

NUCLEUS. A head or center. The nucleus of a comet is the bright point from which the tail streams.

Occult. To cover from view or conceal. In an eclipse, one heavenly body occults another. The Moon continually occults the stars behind it.

Opposition. The point at which two heavenly bodies are furthest removed from each other, or their longitudes differ by 180 degrees. Thus there is an opposition of the Sun and Moon at every full moon.

Optical Doubles. Two stars which appear double to the unaided eye because they are in the same angle of vision, though perhaps many millions of miles apart.

Orbit. A path or track. In astronomy it alludes to the path in space along which a heavenly body travels.

Outer Planets. The four planets furthest removed from the Sun—Jupiter, Saturn, Uranus, Neptune.

Parabola. A curve commonly described as the intersection of a cone with a plane parallel with its side; the two sides of which constantly diverge.

Parallax. The difference between the apparent position of a heavenly body if viewed from the Earth's center and from a point on its surface.

Parallels. Lines lying in the same plane which remain the same distance from each other.

Perihelion. The point in the orbit of a planet or other body when it is nearest the sun. The reverse of *aphelion.*

Phases of the Moon. The aspects or changes which the Moon assumes, beginning as a narrow crescent, proceeding to full, and waning to another crescent.

Photosphere. The radiant surface, which is visible to Earth inhabitants, of the Sun; also of fixed stars.

Planets. Specifically applied to the eight members of our solar system, bodies which revolve around the Sun and receive their light and warmth therefrom. These are: Mercury, Venus, Earth, Mars, Jupiter, Saturn, Uranus, and Neptune. Besides these there are many small planets or planetoids.

PLANETESIMAL HYPOTHESIS. Opposed to the nebular hypothesis, this theory is that our solar system was created from a central sun surrounded by planetoids and nebula, from which were successively created our Sun and its satellites.

PLANETOIDS. Minor planets or asteroids. Several hundred such bodies have been discovered in our solar system, lying chiefly in the vast field between Mars and Jupiter.

POLAR. Pertaining to a pole, or the extremity of an imaginary line or axis passing through the Earth or other celestial body.

PRECESSION OF THE EQUINOXES. A slow motion of points of the equinox on the ecliptic from east to west, at the rate of about fifty degrees annually—thus changing the time at which constellations are visible.

PROMINENCES. Flashes of light or eruptions of gas above the chromosphere of the Sun. These are beautifully colored, fantastically shaped, and often thousands of miles high.

PTOLEMAIC THEORY. Ptolemy, an Alexandrian astronomer, published about 140 A. D., an "Almagest," in which he assumed that the Earth was the center of our system, and around it the other heavenly bodies revolved.

QUARTER MOON. One of the four phases through which the Moon passes in its monthly waxing and waning.

RETROGRADE. The backward movement which the Earth appears to have, relative to the fixed stars.

RILLS. Deep and wide crevices which are visible upon the surface of the Moon.

SATELLITE. A smaller body which revolves around or is

attracted by a larger one. The best example is the Moon, which is the Earth's satellite.

SHIFT. The apparent displacement of a star on the background of the sky, when viewed from different angles.

SHOOTING STARS. Popular name given to meteors, or wandering bodies which strike our atmosphere and become luminous.

SIDEREAL DAY. The exact time required by the Earth to make a complete turn on its axis, as measured by the stars.

SIDEREAL PERIOD. Time as measured by the apparent motion of the stars, as opposed to that measured by the Sun.

SIGNS OF THE ZODIAC. The names given to twelve constellations on account of their fancied resemblance to certain figures. See Zodiac.

SOLAR CYCLE. A seven-year period during which the spots on the Sun increase and decrease.

SOLSTICE. The time of year when the Sun pauses and then turns back on its apparent course—usually on June 21st, which is the longest day, and December 22nd, which is the shortest day in northern latitudes.

SPECTROSCOPE. An optical instrument for determining and measuring the various rays emitted by the Sun and other luminous bodies.

SPECTRUM. An image formed by rays of light, and usually seen as parallel bands of varying colors.

SPRING TIDE. The high tide which occurs twice each month due to the combined attraction of the Sun and Moon.

STAR. Name given to distant celestial bodies, many of which are larger than our Sun, the latter being itself a star.

Star Cluster. An apparent close grouping of star masses partly due to their revolution around a common center, and partly the result of their great distance from the eye.

Sun. Name given to any glowing and light-emitting heavenly body. In addition to the Sun which is the center of our solar system, there are many other suns among the fixed stars.

Synodic Period. The interval of time which elapses between the conjunctions of two heavenly bodies.

Tail (of Comet.) A nebulous mass extending from the head of a comet, always in a direction away from the Sun.

Temporary Stars. Stars which wax and wane and sometimes disappear from view altogether. Thought to be due to friction caused by contact with the nebulous masses.

Transit. The act of passing across a field of vision, as for example the transit of Venus across the face of the Sun.

Uranolith. A meteorite; a small wandering heavenly body which shatters on encountering the Earth's atmosphere.

Variable Star. A star which varies in light-giving properties, due it is thought to obscuration by another star, or explosions of gas.

Vernal Equinox. The period in the spring when the Sun crosses the Equator and the days and nights are of equal length. Usually March 21st.

Year of Time. The time required for a given planet to make one revolution around the Sun.

Zenith. The point in the celestial sphere that is exactly overhead, or at an angle of ninety degrees from the Earth's plane.

ZODIAC. An imaginary zone or girdle stretching around the celestial sphere and extending eight degrees on each side of the ecliptic. It is the path followed by the Moon and all of the planets across the sky. It is divided into twelve parts, one for each month, and marked by a certain constellation, called a "sign of the zodiac."

ZODIACAL LIGHT. A disk of faint light surrounding the Sun, lying near the plane of the ecliptic and extending beyond the Earth's orbit.

INDEX

Spotted something that isn't quite right?

Let us know and we'll fix it for future reprints.
And as a thank you for your diligence, we'll
send you a free Sothis book of your choice.

publisher@sothis.press

This page intentionally left blank

Made in United States
Orlando, FL
05 November 2026

90229005R00194